Men-at-Arms • 364

The Russian Army 1914–18

Nik Cornish • Illustrated by Andrei Karachtchouk
Series editor Martin Windrow

First published in Great Britain in 2001 by Osprey Publishing,
Elms Court, Chapel Way, Botley, Oxford OX2 9LP, United Kingdom
Email: info@ospreypublishing.com

© 2001 Osprey Publishing Ltd.

All rights reserved. Apart from any fair dealing for the purpose of private study, research, criticism or review, as permitted under the Copyright, Designs and Patents Act, 1988, no part of this publication may be reproduced, stored in a retrieval system, or transmitted in any form or by any means, electronic, electrical, chemical, mechanical, optical, photocopying, recording or otherwise, without the prior written permission of the copyright owner. Enquiries should be addressed to the Publishers.

ISBN 1 84176 303 9

Editor: Martin Windrow
Design: Alan Hamp
Index by Alan Ruttter

Originated by Magnet Harlequin, Uxbridge, UK
Printed in China through World Print Ltd.

01 02 03 04 05   10 9 8 7 6 5 4 3 2 1

FOR A CATALOGUE OF ALL BOOKS PUBLISHED BY
OSPREY MILITARY AND AVIATION PLEASE CONTACT:

The Marketing Manager, Osprey Direct UK
PO Box 140, Wellingborough
Northants, NN8 4ZA, United Kingdom
Email: info@ospreydirect.co.uk

The Marketing Manager, Osprey Direct USA
c/o Motorbooks International
PO Box 1, Osceola, WI 54020-0001, USA
Email: info@ospreydirectusa.com

www.ospreypublishing.com

Dedication

The author dedicates this book to Elizabeth,
and to his children James, Alex and Charlotte.

The artist dedicates his paintings for this book to the memory of Vladimir Zweguintzow, a Russian patriot, who devoted his life to popularising the history of the Russian Imperial Army.

Acknowledgements

The author wishes to extend sincere thanks for their assistance to Dmitry Belanovski, Stephen Connolly, Andrei Simonov and Stephen Perry; and to Col.A.K.Nikonov, Director of the Central Museum of the Armed Forces, Moscow, and to his staff, particularly Ms H.A.Gmyrya.

The artist wishes to acknowledge the kind help of A.Deryabin, M.Khvostov, S.Laptev, R.Palasios-Fernandes and A.Valkovich.

All photographs not specifically credited otherwise are the property of STAVKA Military Research. The credit (CMAF) indicates images kindly supplied by the Central Museum of the Armed Forces, Moscow.

THE RUSSIAN ARMY 1914–18

INTRODUCTION

This relaxed image of an infantryman in 1914 clearly shows the summer shirt-tunic, service cap, water bottle and mess tin. Lacking only ammunition pouches to complete his field kit, Pte.Pavel Zherdev has his greatcoat rolled across his chest in the distinctive Russian manner. During Stalin's purges of the 1930s Zherdev, by then the First Commissar of a region on the Volga, was executed.

THE RUSSIAN ARMY of the First World War has for decades suffered an image problem compounded by politics, secrecy and ignorance. The memoirs of those commanders who survived to write them often tend to be apologist or self-seeking. Russia's withdrawal from the war is blamed upon politicians of varying shades of opinion. Shell shortages, lack of Western support, traitors in high places, Russia's sacrifices in the interests of France in 1914 and Italy in 1916 – all these factors have credibility, but none tells the whole story.

Many Western historians tended to fall in with one or other of these schools of thought, until the publication of Professor Norman Stone's *The Eastern Front 1914–17* in 1975. Stone demonstrates that by late 1916 Russia was producing sufficient munitions; but that her inability to adapt to wartime imperatives such as feeding the urban population and developing a viable supply system led to her collapse into revolution.

* * *

At the beginning of the 20th century the Russian empire covered eight million square miles, with a population of some 170,000,000 people, and was ruled over by one man: Tsar Nicholas II, whose Romanov dynasty had celebrated 300 years of power in 1913. The authority of the Tsar was absolute but, as the revolution of 1905–06 had shown, it rested on the support of the army. Japan's victory in the Russo-Japanese War (1904–05) led to a review of Russia's armed forces during the years leading up to 1914. When Gen.V.A.Sukhomlinov was appointed Minister of War in March 1909, reform became a matter of priority.

It became clear that reform of the armed forces and industrialisation would have to proceed together. Domestic production of small arms and field artillery was sufficient, but for heavier artillery, communications equipment and other modern necessities it was woefully inadequate. It was necessary to import these items until Russian industry could produce what was required. The period 1910–14 saw change on a scale unprecedented during peacetime: rates of pay were increased to encourage the retention of experienced men, hundreds of officers were retired as incompetent, conscription was expanded to create a larger reserve pool, and the military budget was increased.

Inevitably there was opposition to these reforms, which polarised into hostility between those who supported Sukhomlinov's modernisation programme and the more traditionalist adherents of the Grand Duke Nicholas, uncle of the Tsar, Commander of the Imperial Guard and the St Petersburg Military District. Consequently reform was implemented only slowly; and in the matter of artillery it was complicated by strategic as well as industrial problems.

Russia's western defences were based on the assumption of an invasion from Germany or Austria-Hungary. Mobilisation at the outbreak of war would be slow, due to Russia's vast size and under-developed rail system. To buy time for mobilisation a line of colossal fortresses, bristling with artillery, was built during the latter part of the 19th century at key points throughout Russian Poland. However, the range and power of 20th century field artillery outclassed the fortress artillery, which required up-grading. During the five years before 1914 a large percentage of the artillery budget was invested in modernising the fortress guns – at the expense of the mobile heavy artillery of which Russia was particularly in need.

However, war seemed a long way off; and the 'Great Programme' of modernisation approved in 1914 was due for completion in mid-1917. Russia and France had been allies since 1893, and through the Anglo-French treaty this also linked Russia to Britain. A French loan was arranged, specifically to construct railways in Poland to speed up mobilisation. It was anticipated that when all the pieces of the Great Programme were in place the armed forces of Russia would be prepared for any scale of international conflict.

The strategic situation

Both Germany and Austria-Hungary were, by early 1914, extremely concerned about the modernisation of Russia's forces. The necessity for action before Russia's investments bore fruit was becoming critical, as Germany's plan for European war rested on the prerequisite of Russia's mobilisation being slower than Germany's conquest of France. Germany planned to commit the greater part of her army in the West to over-running France, leaving some two army corps and local forces to defend her eastern border. These troops, combined with the Austro-Hungarian armies, were thought to be sufficient to hold Russia until victory in the West released the bulk of Germany's forces to turn against her.

Reality dictated that Russia would face Germany and Austria-Hungary to the west and the Ottoman Turks to the south; it was therefore essential to decide where to place the main weight of the Russian army during the mobilisation period. The Army of the Caucasus was deemed capable of dealing with the Ottoman threat; thus it was a choice between Germany and Austria-Hungary. Clearly Austria-Hungary was temptingly the weaker, but Germany posed the greater threat. Two plans had been drawn up: Plan 19 gave greater weight to an offensive into East Prussia, and Plan 19 Revised, drawn up in May 1912, reduced the forces committed to East Prussia, stressing Austria-Hungary as the main target.

Plan 19 Revised necessitated the creation, at the outbreak of war, of two 'Fronts', one to command each operation – the North-Western and the South-Western, both to be overseen by a supreme headquarters known as STAVKA. The Austria/Germany dilemma was one that the Russians never fully resolved. To further complicate the issue the Russians had given an assurance to their French partners that they would launch an offensive into East Prussia during the early weeks of any war.

Following the assassination of the heir to the Austro-Hungarian throne in June 1914 and the intense political activity that ensued throughout Europe, Russia's mobilisation was carried out according to

Tsar Nicholas II appointed his uncle the Grand Duke Nicholas Nicholaevitch Supreme Commander of the Armed Forces of Russia on 2 August 1914. The Grand Duke is pictured here at STAVKA – Supreme Headquarters – at Baranovitchi during 1915. Effective command was in the hands of Nicholas' various chiefs-of-staff.

Plan 19 Revised, resulting in the invasion of both East Prussia and Austria-Hungary. Despite the distances involved and the incomplete rail network near the western borders the mobilisation was carried out with remarkable efficiency, much to the horror of the Central Powers.

Unsurprisingly, the weather was to play an important part in the war on the Eastern Front. The severity of the winters and the incredible mud generated by the thaws limited the campaigning season to the period May–October. The sheer scale of this front dominated the thinking of the Central Powers, as they did not wish to repeat Napoleon's failure by advancing too far into Russia. The plains of Poland ended at the Carpathian Mountains to the south, providing a natural defence for Austria-Hungary. To the east of the Carpathians lay the endless steppes of the Ukraine and the almost impassable Pripyat Marshes. Russia's Baltic provinces, bordering East Prussia, were scantily developed, flat and largely featureless, but provided the shortest route to the capital Petrograd (as St Petersburg was renamed at the outbreak of war), and the bases of the Baltic Fleet.

CHRONOLOGY NB: *The Western calendar is used.*

1914:

1 August Germany declares war on Russia; mobilisation gathers momentum. **2 August** Grand Duke Nicholas Nicholaevitch appointed Supreme Commander-in-Chief with Yanushkevitch as Chief of Staff and Danilov as Quartermaster General. STAVKA established at railway junction of Baranovitchi. **7 August** Russian 9th Army begins to assemble at Warsaw to invade Silesia. **15 August** Russian 1st Army crosses into East Prussia from the east. **20 August** Russian 2nd Army crosses into East Prussia from the south. 1st Army defeats German 1st Corps at Gumbinnen; German commander panics and is replaced by Hindenburg and Ludendorff. Russian 3rd Army crosses into Austria-Hungary. Austrian and Russian cavalry clash at Jaroslawice. **21 August** Russian 8th Army crosses into Austria-Hungary. **23–24 August** Russian 4th Army defeated by Austrians at Krasnik in southern Poland; 9th Army moves south; Silesian invasion postponed. **25–30 August: Battle of Tannenberg** Russian 2nd Army encircled and destroyed in East Prussia. **26–28 August** Austro-Hungarian 3rd Army defeated on River Zlota Lipa. **29–30 August** Austro-Hungarian 4th Army

This painting of the 3rd Squadron of the Horse Guards capturing a German battery on 19 August 1914 quite accurately reproduces the uniforms, the tactics and the elan of this elite unit. The cap and shirt-tunic are light summer khaki, the breeches are blue-grey striped with red, the cross belts white and the shoulder straps black piped with red. All non-Cossack cavalry slung their carbines over the left shoulder; half of each regiment carried the lance and in action formed the front rank. Bayonets were issued for dismounted fighting. The Horse Guards rode black horses and used the standard dark brown horse furniture.

defeated at Gnila Lipa. **30–31 August** Russian 4th and 5th Armies defeated at Zamosc-Komarow.

3 September Lemberg (Lvov) falls to Russian 3rd Army. Situation on Austro-Hungarian eastern flank becomes critical; to the north their advance towards Lublin stalls. **11 September** Austro-Hungarian forces retreat towards Przemysl; bad weather slows Russian pursuit. Germans send troops to support Austria-Hungary. **Mid-September: Battle of the Masurian Lakes** in East Prussia; defeat and withdrawal of Russian 1st Army. **16 September** First siege of Przemysl begins. **25 September** German advance following Masurian Lakes victory is checked.

October German advance on Warsaw begins; Russian counter-attacks succeed, Germans begin to retreat on **20 October**. At the same time Austro-Hungarians advance across River San, but by **26 October** they are defeated and driven back. **14 October** Siege of Przemysl raised. **Late October** By now Russia has deployed 82 infantry divisions against the Central Powers, with a further 16 defending Baltic and Black Sea coasts. **November** Declaration of war on Turkey; Caucasian Front prepares for Turkish invasion. **11 November** Russia's planned invasion of East Prussia, betrayed by German breaking of wireless codes, is pre-empted by German attacks, but these are held. Invasion of Silesia indefinitely postponed. First serious discussions at STAVKA regarding withdrawal from Poland. **12 November** Russian SW Front defeats Austro-Hungarians in Carpathian Mountains; advances on Cracow but is held on River Dunajec. Przemysl, with c120,000 Austro-Hungarian troops inside, is besieged again.

Early December Lodz falls to Germans, who are again held before Warsaw. First entrenching begins as weather deteriorates. Austro-Hungarian attempt to relieve Przemysl fails. Turkish invasion of the Caucasus ends in disaster at battle of Sakrimash.

Trench warfare: a forward artillery observation post. The officer viewing the enemy lines through stereoscopic binoculars passes the information to his colleagues who relay the information to the gun position by field telephone. The landlines were naturally vulnerable to artillery fire, but were more secure – and available – than wireless sets. The telephone equipment was usually manufactured abroad.

1915:

Early January First recorded use of gas at Bolimov in Poland (effects negligible). **23 January** Austro-Hungarians recapture Carpathian passes.

7 February German attack from East Prussia develops into Second Battle of the Masurian Lakes; Russian 10th Army destroyed. Germans begin siege of Osowiec, which is raised after a month. Russian counter-offensive in Carpathians and along the Dniester River pushes Austro-Hungarians back in snowstorms.

Early March Counter-offensive by remains of Russian 10th & 12th armies drives Germans back into East Prussia. NW Front stabilises but its commander, Gen.Ruzski, resigns as he is allowed neither to invade East Prussia nor to withdraw from Poland. **22 March** Przemysl falls to Russians who take over 100,000 prisoners. **End March** Gen.M.V.Alexeyev replaces Ruzski and adopts his ideas.

April Continuing success of Ivanov's SW Front renders Austria-Hungary's situation critical. To relieve pressure Germans advance against thinly defended Courland (western Latvia). **10 April** Ivanov halts SW Front's progress and calls for reinforcements.

2 May Germans open major offensive between Gorlice and Tarnow with (for the Eastern Front) unprecedented weight of artillery support. Brunt of German attack directed at Russian 3rd Army – short of artillery ammunition and poorly entrenched. **10 May** After losing nearly 200,000 men and 140 guns, 3rd Army is given permission to retire to River San. **16–19 May** Austro-German troops attack River San positions. Russian 9th Army attacks and overruns much of the Bukovina. **23 May** Italy joins the Allies (this does not immediately affect Austria-Hungary's strategy). **20–25 May** Russians hold on the San, but 9th Army retreats to River Dniester.

4 June Przemysl recaptured by Austro-Hungarians. Gen.Alexeyev ordered by STAVKA to defend Courland; German advance stopped. **Mid-June** Central Powers resume offensive which, because of artillery tactics employed, is known as 'Mackenson's wedge'. Russian Minister of War Gen.V.A.Sukhomlinov arrested for 'treasonable negligence' and replaced by Gen.A.A.Polivanov. **20 June** STAVKA orders retreat from Galicia. **22 June** Lemberg recaptured by Austro-Hungarians. The six-week Gorlice-Tarnow offensive has cost Russia c300,000 men and 224 guns. STAVKA decides to defend Polish fortress line.

13 July German attacks in Courland and northern Poland make slow progress for heavy casualties. **15 July** Premature Austro-Hungarian advance beaten at second battle of Krasnik. Third battle of Krasnik brings Austrian success; Lublin and Cholm fall by end of July. **19 July** Gen.Alexeyev given permission to abandon Warsaw, and on the 22nd the Russians begin to retire from both northern and southern Poland.

5 August Warsaw occupied by Central Powers. **7 August** With Osowiec fortress as the hinge, Russian forces retire towards Brest-Litovsk, pursued by Germans under Falkenhayn. Accompanied by hundreds of thousands of refugees the Russians retreat, 'scorching the earth' as they do so. Appeals for help to Western Allies are ignored. **15 August** Under political pressure to defend Baltic coastline and route to Russia's capital Petrograd, Guards Army is moved to defend this area from seaborne attack. **17 August** STAVKA divides war zone into three Fronts: Northern, North-Western and South-Western. **18 August** Kovno fortress falls to Germans with loss of 1,300 guns. **20 August** Novo-Georgievsk fortress falls with loss of 1,680 guns. **21 August** Osowiec and Kovel fall. STAVKA relocates to Mogilev. **24 August** Brest-

The realities of war: wounded men are being helped aboard a hospital train. The battered appearance of their dress is a far cry from the studio portraits of 1914. The medical attendant on the train is wearing a fleece cap with cockade; the side flaps could be let down to protect the ears – cf Plate D3. The second and fourth casualties from the left are wearing peakless field caps, the third a black fleece cap, possibly of natural wool. The majority of medical personnel wore Red Cross brassards. Wealthy individuals privately supported many hospital trains.

Litovsk fortress abandoned. **End August** Austro-Hungarian 'Black-Yellow' offensive takes railway junction of Lutsk.
1 September The Tsar takes the role of Supreme Commander-in-Chief and appoints Grand Duke Nicholas to the Viceroyalty of the Caucasus. Gen.Alexeyev becomes Tsar's Chief of Staff, and effective commander of Russian armed forces. **18 September** Vilna falls to Germans. **22 September** 'Black-Yellow' offensive halted by Gen.Brusilov's Russian 8th Army; SW Front stabilises. **26 September** On North and NW Fronts Russian resistance hardens and German attacks are called off. Both sides entrench as Central Powers consolidate their gains.
October Russians lose the first battle of Lake Narotch. **Mid-October** Central Powers, aided by Bulgaria, invade Serbia. Russians reorganise. The length of the Eastern Front has been cut from 1,700 to 1,000 miles; domestic production of munitions is growing and supply is improving. 1,000,000 refugees have moved into the Russian heartland, as have strategic industrial concerns.
November STAVKA discusses possible seaborne invasion of Bulgaria. When the admirals refuse, Gen.Alexeyev threatens to turn Black Sea Fleet into 'an infantry brigade'. His threat has no effect, and he proceeds to create 'Army of Descent' around core of 7th Army based near Odessa.
December Measures to replace heavy losses among officer corps lead to recruitment of many disaffected men, who are frequently put in charge of depot units in Kiev, Moscow and Petrograd. Russia extracts promises that during 1916 Western Allied efforts will be synchronised and support given at times of crisis. **27 December** Russia launches limited offensive in the Bukovina in support of Serbia. Although a failure, it is carefully analysed by staff officers of SW Front with a view to incorporating lessons into new tactical doctrine.

1916:

January Gen.Alexeyev announces a front line strength of 2,000,000 men.
February Turkish city-fortress of Erzurum falls to Russians. **21 February** Start of German offensive at Verdun, France. **24 February** STAVKA considers Lake Narotch as site of next offensive, to take place in April.
March Trabzon taken by Russians following combined land and sea operations. **18–19 March** Russians attack at Lake Narotch in response to French requests for diversion. Results of bombardment poor due to saturated ground; infantry attack as weather deteriorates. **End March** Narotch operation called off after 120,000 casualties.
April Germans recapture ground lost at Lake Narotch. Gen.A.A.Brusilov becomes commander of SW Front. **14 April** STAVKA decides main offensive of 1916 to be carried out by W Front with subsidiary operations by SW Front.
20 May Success of Austro-Hungarian offensive in Italy leads Italian king to appeal to Tsar for help. Western Front's preparations are incomplete but Brusilov agrees to attack early.

Following the loss of Poland and the advance of the Austro-Germans Tsar Nicholas took personal command. Effectively the power rested with his Chief-of-Staff Gen.M.V.Alexeyev, between September 1915 and May 1917 when he was replaced by Brusilov. Alexeyev came from a very humble background, and his father had risen from serf to officer; a gifted administrator, he had 'only the good of Russia at heart'. During the summer of 1915 STAVKA relocated to Mogilev, where it was to remain until the end of the war.

4 June Start of the **'Brusilov Offensive'**. All four armies of SW Front – 8th, 9th, 7th & 11th – bombard Austro-Hungarian lines; innovative Russian artillery fire plan causes chaos as Russians breach first and second Austro-Hungarian lines. **5 June** 8th Army breaks Austro-Hungarian third line. **6 June** Austro-Hungarian 4th & 7th Armies near to collapse. **9 June** Austro-Hungarian 7th Army retreats into the Bukovina. **12 June** Brusilov halts advance after taking c200,000 prisoners and 216 guns. **24 June** Austro-Hungarian offensive in Italy is called off; Central Powers counter-attacks on River Stokhid fail.

1 July In France, British offensive on the Somme begins. **2 July** Main Russian attacks by W Front begin; by 8 July Russian casualties reach 80,000 for negligible gains. **3 July** Gen.Brusilov attacks again, driving back German Sud Armee.

Mid-July In France, Germans call off Verdun attacks. Gen.Brusilov is given more troops but, due to court intrigue, does not have authority over the Guards Army. **28 July** Start of series of suicidal attacks through marshes and swamps near Kovel, in which Guards Army is decimated.

August Russian 8th Army continues advance westwards, but supply lines grow longer and resistance hardens. 9th Army drives into the Bukovina; 8th and 9th Armies take Halicz. **End August** Romania declares war on Central Powers and invades Transylvania. By this time Central Powers losses stand at over 500,000, and Austro-Hungarian troops are being transferred from Italy. However, Brusilov's 'Broad Front' approach has been discontinued in favour of the 'Narrow Front' as used with such disastrous consequences at Kovel.

September Gen.Alexeyev sends one cavalry and two infantry divisions to support Romania. **6 September** Central Powers move against Romania from the south. **15 September** Romanians move troops from Transylvania to the south.

3 October Russo-Romanian offensive against Bulgaria fails. **6 October** Central Powers re-occupy Transylvania; end of Russian offensive in the Bukovina. **21 October** Central Powers, mainly Bulgarian and Turkish units, drive back Romanian and Russian troops in southern Romania. **End October** Gen.Alexeyev sends Russian reinforcements to Romania; Bulgarian-Turkish advance halted.

November Austro-German troops break through Carpathian passes. **23 November** Central Powers cross the River Danube.

December Romanian counter-attack fails and army falls back into Moldavia. **7 December** Romanian capital Bucharest falls to Central Powers. **End December** Romanian line stabilises along River Siret.

1917:

Early January Russians in *de facto* control of Romanian Army; Romanian Front created. **January** Russian NW Front launches successful operation near Mitau on Baltic Sea. Discussions at STAVKA regarding offensive on SW Front and radical reorganisation of armed forces. Inter-Allied conference reiterates policy of mutual support and promises delivery of munitions, including heavy artillery and aircraft.

High in the mountains on the Caucasus front a group of officers discuss the situation. The landscape varied from bleak and barren to verdant and wooded nearer the Black Sea coast.

February–March Strikes and demonstrations in Petrograd are supported by replacement units in city; military and civilian authorities lose control. Councils known as *Soviets* spring up across Russia, acting as alternative, radical form of leadership.

12 March Provisional Government established in Petrograd; revolution accomplished with little bloodshed. **14 March** Order No.1 issued by the Petrograd Soviet, dramatically reducing powers of officers, influences whole armed forces and leads to breakdown in discipline. **15 March Tsar abdicates** in favour of his brother Mikhail, who rejects the throne. **The Petrograd Soviet calls for an end to the war, and world revolution.**

April Regimental revolutionary committees exert increasing power over military operations and appointments. The front remains quiet as the Central Powers observe the situation. **16 April** V.I.Lenin, leader of the *Bolshevik* party, returns from exile to Petrograd and calls for 'Bread, Peace and Land'. **30 April** To secure financial and military aid from the West, Miliukov, for the Provisional Government, declares intention to continue fighting. (He is forced to resign on 18 May.) Russian activity on the Caucasian Front ceases due to supply crisis.

May The charismatic Alexander Kerenski becomes Minister of War. Fighting again breaks out at the front. Brusilov replaces Alexeyev as Supreme Commander-in-Chief. Russians withdraw in Anatolia.

June Preparations for summer offensive on SW Front stepped up in attempt to emulate success of 'Brusilov Offensive' of previous year.

1 July Start of Russian summer offensive, which becomes known as the **'Kerenski Offensive'**. **July** Under the greatest weight of shells used by Russian artillery during the war the defences of forces facing SW Front crumble. Russian 7th & 11th Armies advance through 30km breach towards Lemberg. Both W & NW Fronts attack in support. However, within a few days morale erodes; when facing anything but negligible opposition troops refuse to attack, or withdraw. **19 July** German counter-offensive; Russian troops fall back, losing cohesion and discipline.

1 August Gen.L.G.Kornilov replaces Brusilov as Supreme Commander-in-Chief with intention of restoring discipline.

August The front line stabilises as Central Powers encounter firmer resistance, particularly on Romanian Front.

3 September Russian 12th Army abandons Riga on Baltic coast. **7–9 September** Failure of 'Kornilov's Coup', bungled attempt to overthrow Provisional Government; Kornilov imprisoned with other senior officers. **14 September** Russia declares itself a republic.

12–19 October 'Operation Albion' – Germans take control of Gulf of Riga by capturing Moon and Oesel islands.

Another photograph from the Caucasus, though it could be anywhere on the Eastern Front: a group of troops, dressed for the bitter weather, brew tea. Greatcoats and fleece caps are clearly essential in these conditions.

7 November Lenin leads **Bolshevik-inspired coup in Petrograd** which ousts the Provisional Government. **Mid-November** Kerenski fails to rally support for Provisional Government, and goes into exile. **18 November** Declaration of Soldier's Rights includes abolition of ranks and election of commanders. **20 November** Trotski, Commissar for Foreign Affairs, formally notifies Allied Powers of change of government and Decree of Peace. **21 November** Fraternisation on all fronts is authorised by Krylenko, Commissar for War. **23 November** Gradual demobilisation of the army is declared. **26 November** Trotski formally approaches Central Powers for an armistice.

3 December Gen.N.Dukhonin, last Supreme Commander-in Chief of Russian Armed Forces, is murdered at Moghilev by Red Guards. With authority collapsing, the army begins to drift home. **15 December** Russia signs 30-day armistice with Central Powers.

1918:
January Peace talks with Central Powers begin at Brest Litovsk. **28 January** Formation of Red Army announced.
18 February Frustrated by Russian negotiators, Central Powers advance along entire Eastern Front, occupying vast tracts of western Russia.
3 March Russia signs Treaty of Brest Litovsk and withdraws from the war.

ORGANISATION OF THE ARMY

In peacetime the empire was divided into 12 military districts, each under a commander-in-chief: St Petersburg, Vilna, Warsaw, Kiev, Odessa, Moscow, Kazan, the Caucasus, Turkestan, Omsk, Irkutsk, and the Pri-Amur. Russian land forces consisted of the Standing Army and the Imperial Militia (*Opolchenie*). The Standing Army comprised the regular army and its reserve; the Cossacks; and the 'Alien' troops (*Inorodtsi*) – these latter being non-Slavic Imperial subjects. The strength of the armed forces immediately prior to mobilisation was officially estimated at 1,423,000 men; after mobilisation was complete this would increase to some 5,000,000 – the Russian 'steamroller' which was relied upon to crush any opponent by sheer weight of numbers.

Liability for conscription began at the age of 21 and lasted until 43. The first three years (for the infantry and artillery) or four years (for other branches of service) were served with the colours; the next seven years were spent in the first class reserves, and the final eight in the second class. Men could also volunteer, in which case their conditions of service were privileged. Recruitment was in the main from the Russian, Christian population of the empire, thus excluding Moslems, who paid a tax in lieu. (Units referred to as 'Finnish' were Russian formations based in Finland, as the Finns themselves were exempted from conscription.) Generally 50 per cent of those called up in any year were exempted on physical, personal, economic or educational grounds; and retention of experienced men who could provide the NCO class was poor.

The image of the *frontovik*, the typical front line soldier; he has a worn greatcoat and ragged *bashlyk* (cowl) wrapped across the chest and tucked into the belt, with the hood hanging down the back. The cowl was issued to all ranks and branches of service; many units piped the seams in regimental colours. Beards were permitted in most units. The shoulder straps are plain khaki. The service cap seems to be the parade version, which would have been dark green, with a red band and red crown piping for the first regiment in each infantry division; the others had blue, white and green bands respectively.

Cossacks served from 20 until 38, doing 12 years in the 'field class' – four years each in the first, second and third category regiments – and the remaining time in the reserve. The 'Alien' troops were volunteer irregular cavalry units recruited from Moslem tribes.

The Imperial Militia comprised the majority of the men granted exemption from the regular army between the ages of 21 and 43. The first category was used to strengthen or complete the Standing Army and was divided according to age. The second, less physically capable group, was used to form non-combatant units. Provision existed for the organisation of 640 battalions (*druzhina*) of militia. During the years preceding the revolution of March 1917 several million militiamen of all classes were mobilised.

Infantry

The 208 recruiting areas in Russia provided men for the 208 line infantry regiments. The Guard, Grenadiers, Rifles, artillery, cavalry, and engineers drew recruits from any district. The minimum height requirement was 5 feet and half an inch.

The field army was divided into 37 army corps: the Guard, the Grenadier, I to XXV, I to III Caucasian, I and II Turkestan, and I to V Siberian. These included all the infantry divisions with their attached artillery. The normal composition of an army corps was two infantry divisions, one *divizion* of light howitzers (two 6-gun batteries), and a battalion of sappers; the term *divizion* refers here to a half-regiment. An infantry division consisted of four regiments each of four battalions, and a field artillery brigade of six 8-gun batteries.

Of the 236 infantry regiments 12 were Guards and 16 Grenadiers. The Guards regiments were named, the Grenadiers numbered 1–16, and the Line named and numbered. The 4th or Caucasian Grenadier Division was permanently stationed in the Caucasus.

An infantry regiment was made up of four battalions each of four companies, plus a non-combatant company. Infantry regiments were allotted serially to divisions; thus the 17th Infantry Division would include the 65th to 68th Regiments inclusively. A wartime company would number 240 men and four or five officers. At regimental level machine gun, communications and scouting detachments, known as *kommandos*, brought total regimental strength up to about 4,000 men. In 1914 each regiment had eight machine guns; 14 mounted messengers and 21 telephonists in the communications detachment; and 64 specially trained scouts including four cyclists.

Infantry regiments from Siberia and Turkestan were known as Rifles, but organised like other infantry regiments with four battalions. The Rifle regiments

Cossacks crossing a river in East Prussia in summer 1914 – not a method to be used under fire, but one that is obviously effective for men who were raised in the saddle from infancy. Support for the rider was given by crossing the stirrups over the saddle. Cossack horses were renowned for their stamina and ability to live on short rations; they were specially bred on government stud farms out on the steppe. The Cossacks were regarded as irregular troops, which allowed them a degree of freedom not available in the Guard and Line cavalry (e.g. the quiff or 'lovelock' hairstyle favoured by the Don Cossacks, as in Plate B3).

'proper' had only two battalions each. There were four Guards Rifle regiments forming the Guards Rifle Brigade; 20 Line Rifle regiments numbered 1–20, forming the 1st to 5th Rifle Brigades; 12 'Finnish' Rifle regiments numbered 1–12, forming the 1st to 3rd Finland Rifle Brigades; and eight Caucasian Rifle regiments numbered 1–8, forming the 1st and 2nd Caucasian Rifle Brigades. The 22 Turkestan Rifle regiments were numbered 1–22; the 1st to 4th Turkestan Rifle Brigades had four battalions each, the 5th and 6th three. Each Rifle brigade contained a Rifle Artillery Brigade of three 8-gun batteries. By 1914 the title Rifle signified nothing more than the unit's historical role.

On mobilisation, 35 reserve infantry divisions were formed numbered 53rd to 84th, and 12th to 14th Siberian. The establishment of these divisions was identical to that of the regulars, though their artillery was often equipped with less modern ordnance.

Cossack infantry were known as *plastuni*. Initially only the Kuban Cossack Host raised infantry, but the practice later spread to others. They were organised into brigades each of six battalions, without artillery. In 1914 three brigades were assigned to the Caucasian Front.

Cavalry

Russia had the largest cavalry establishment of all the belligerent nations in 1914. There were four groups: the Guard (see below under 'Elite Units'), the Line, the Cossacks and the Alien troops. Line cavalry and Cossack regiments were all composed of six squadrons, giving a combat strength of about 850 men; a Cossack squadron was known as a *sotnia*. Although the historical lancer, hussar and dragoon titles were retained they had no tactical significance. Specialist detachments for scouting, communications and demolition were as per the infantry. Attached to each division was a machine gun detachment of eight guns.

In 1914 there were 20 dragoon, 17 lancer and 18 hussar regiments of the line. On mobilisation 24 cavalry and Cossack divisions were formed, and an additional 11 independent cavalry and Cossack brigades. Cavalry divisions were formed of two brigades, the first grouping a dragoon and a lancer regiment, the second a hussar and a Cossack regiment. The regiments were organised serially; thus the 3rd Cavalry Division would include the 3rd Dragoon, Lancer and Hussar Regiments. There was also a dragoon division with the Army of the Caucasus, the Caucasian Cavalry Division.

Cossacks

The Cossacks were divided into two broad groups: those of the steppe, *Stepnoy*, and those of the Caucasus, *Kavkas*. The Caucasians split into two *voiskos* ('Hosts'), the Kuban and the Terek; the Steppe, into the Don,

The crew of this M1902 Putilov 7.62cm field gun wear the dark green pre-war breeches striped red, with summer khaki service cap and shirt-tunic. The shoulder straps are worn coloured side up, showing scarlet with crossed cannon barrels stencilled in yellow.

While the field artillery was adequate, the lack of heavier artillery in the field armies was serious. At the outbreak of war some 2,813 modern guns plus 3,000 older ones were confined to the fortresses, while the army in the field had only 240 heavy guns and howitzers between them.

This brother and sister are pictured in January 1916. The lady holds the rank of *praporshik* (ensign) in the 9th Siberian Rifles, her brother that of captain in the same regiment; he wears an award of the Cross of St Stanislaus. Female soldiers were not common but were an accepted part of military life. Note that his 'French'-style tunic has a stand-and-fall collar, while she wears a *gymnastiorka* buttoning the female way, right to left.

Siberia, Orenburg, Ural, Astrakhan, Trans-Baikal, Semiretchi, Amur and Ussurski *voiskos*. The largest Host was that of the Don.

The first category regiments were maintained in peacetime, the second and third categories being activated when the need arose. The Don Host raised 54 regiments, the Kuban 33, the Orenburg 16, and the others in proportion to their populations. Cossack divisions were created in the main from a single Host, but Combined Divisions formed from different Hosts were raised during the war. Second category squadrons were allotted to infantry divisions for use as escorts, messengers, local security troops, etc. Some 50 batteries of Cossack horse artillery were raised, mainly from the Don Host.

The **Alien** cavalry were volunteers: the Daghestan Native Cavalry Regiment, the Ossetian Cavalry *Divizion* (half-regiment), and the Turcoman Cavalry *Divizion*. The first two were Moslem tribesmen from the Caucasus, the latter from the Tekin tribe of Turkestan. During August 1914 it was decided to raise a new cavalry division of six regiments among the Moslem peoples of the Caucasus; this Caucasian Native Cavalry Division was nicknamed the 'Savage' or 'Wild' Division, and earned a high reputation (see below under 'Elite Units').

Artillery

The artillery was divided into specialist types: field and mountain; horse and horse-mountain; field howitzer and heavy.

Field artillery was grouped into brigades of two *divizions* each with three 8-gun batteries. A brigade was allocated to each infantry division; this gave three Guard brigades, four Grenadier (1–3 and the Caucasian), 52 line brigades, 11 Siberian Rifle, five Rifle, three Finnish Rifle, two Caucasian Rifle and six Turkestan Rifle brigades.

Mountain batteries were distributed to the Caucasus, Siberia, Finland, Turkestan and Kiev for use in the Carpathian Mountains. The mountain guns could be drawn or broken down for pack transportation.

Horse and horse-mountain artillery was organised into *divizions* of two 6-gun batteries, other than in the Guard, and were attached one to each cavalry division. The three horse-mountain units were distributed one each to the Caucasus (Caucasian Cavalry Division), Siberia (Ussuri Cavalry Brigade) and Kiev (IXth Army Corps).

There were 35 *divizions* of field howitzers (*mortirnie*) each of two 6-gun batteries. One each was allotted to the Guard, the Grenadiers, each of the 25 numbered army corps, 1st–3rd Caucasian, 1st–5th Siberian, and one battery with the 1st Turkestan Rifle Artillery *Divizion*.

What heavy artillery there was was organised into seven *divizions* each of three 6-gun batteries. *Divizions* 1–5 were stationed in the west and the Siberian 1st and 2nd in the east. The first and second batteries in each unit were equipped with 6in. howitzers and the third with 4.2in. guns.

The establishment of fortress artillery was determined by the size of the fortress and the guns it mounted. A large base such as Vladivostock had two brigades, a smaller one a single company.

Technical branches included sapper, railway, and pontoon battalions, field and siege engineer parks, and wireless telegraph companies. The field engineers numbered 39 battalions, one for each army corps and two extra for the Siberian establishment. Other than the Guard engineer battalion (which had four), each battalion had three companies – one or two telegraph companies and a searchlight section.

Wartime innovations

The experience gained during the war caused organisational changes throughout the armed forces. Across the board technical equipment such as telephones became common at all levels. The number of machine guns was hugely increased by imports, captures and expanded domestic production, to the point where MG units organised themselves at almost local level.

Reform of the cavalry divisions was undertaken at the beginning of 1916, when each had attached to it an infantry battalion of three dismounted squadrons. Later in 1916 a reduction in the mounted strength of each cavalry and Cossack regiment from six to four squadrons was approved. The dismounted men were used to increase the infantry element to a three-battalion regiment. Artillery support for the cavalry was to be increased by the provision of howitzer battalions of eight guns – several hundred British 4.5in. howitzers were imported during that year with the promise of more to follow.

During winter 1916/17 STAVKA began reorganising the infantry, reducing the divisional establishment from 16 to 12 battalions and using the surplus to create some 60 new divisions, which were attached to existing corps as a third division. However, the major problem was shortage of divisional artillery; and to overcome this it was decided to reduce the numbers of field guns in passive areas and allocate the surplus to the new divisions. The artillery was to be provided with more heavy guns and these were to be assembled into a 46th Corps, to be known as TAON – the Cyrillic initials for Heavy Artillery on Special Duties. TAON was to be at the disposal of the Supreme Commander-in-Chief; equipment consisted of weapons of varying calibres, including many from France and Britain, which had agreed to give priority to their supply. These were scheduled to arrive during the early months of 1917.

ELITE UNITS

Guards Corps/Special Army

At the outbreak of the war the elite force of the army was the Guards Corps, a self-contained formation including units of all branches, with their own jealously guarded traditions. (Not all of these were based on their fighting record over 300 years – the Pavlovski Guards Regiment was notable for its recruitment of men with snub noses...)

The Guards infantry were divided into three divisions each of four regiments brigaded in pairs, the Guards Rifle Brigade of the 1st to 4th Regiments.

1st Guards Infantry Division
1st Brigade Preobrazhenski & Semenovski Guards Regts;
2nd Brigade Izmailovski & Egerski Guards Regts
2nd Guards Infantry Division
1st Brigade Moskovski & Grenadier Guards Regts;
2nd Brigade Pavlovski & Finlandski Guards Regts
3rd Guards Infantry Division
1st Brigade Litovski & Kexholmski Guards Regts;
2nd Brigade St Petersburg (Petrograd, 1914) & Volynski Guards Regts.

The two Guards cavalry divisions each had three brigades. The Guards cavalry regiments were organised in four squadrons each of 150 men, other than the Horse Grenadiers and the Cossacks, which had six squadrons.

1st Guards Cavalry Division
1st Brigade Chevalier Guards Regt; Horse Guards Regt. *2nd Brigade* His Majesty's Cuirassier Guards Regt; Her Majesty's Cuirassier Guards Regt. *3rd Brigade* His Majesty's Cossack Guards Regt; The Ataman's, HH the Tsarevitch's Cossack Guards Regt (both these regiments recruited from the Don Host); Combined Cossack Guards Regt (this unit was drawn from all the smaller Cossack Hosts in proportion to their populations).

2nd Guards Cavalry Division
1st Brigade Horse-Grenadier Guards Regt; Her Majesty's Lancer Guards Regt
2nd Brigade Dragoon Guards Regt; His Majesty's Hussar Guards Regt
3rd Brigade His Majesty's Lancer Guards Regt; Grodno Hussar Guards Regt.

His Majesty's Personal Escort, the *Konvoi*, recruited four squadrons, two each from the Kuban and the Terek Cossack Hosts.

The 1st to 3rd Guard Artillery Brigades were attached to the appropriate infantry divisions, the Rifle Artillery Brigade to the Guards Rifle Brigade. There were six 6-gun Guard Horse Artillery batteries, the sixth being the Don Cossack Guard Battery. The Guards Howitzer *Divizion* (of two 6-gun batteries) completed the artillery.

The Guard *Equipage* or Crew, naval personnel drawn from the fleet to man the royal vessels, was expanded at the outbreak of war to form two battalions each of two companies; these were given infantry training and sent to the front.

By the summer of 1916 the 1st and 2nd Guards Infantry Divisions had been united to form the 1st Guard Corps, and the 3rd, the Rifles and the *Equipage*, the 2nd Guard Corps. Each corps had an aviation detachment and a heavy artillery *divizion*. These corps were to be known, from 21 July 1916, as the Guards Army. However, following the losses sustained by the Guards during the Brusilov Offensive their numbers were supplemented with line army corps and the whole became in September 1916 the 'Special Army'.

The military muscle behind the March 1917 revolution was provided by the reserve battalions of the Guard in Petrograd. Indeed, the *Equipage*, led by its commander the Grand Duke Cyril Romanov, marched through Petrograd sporting red revolutionary cockades. The

In the summer of 1914 there was one aviation company for each of the 25 army corps, one each for the Guard and Grenadier corps, three Siberian companies and eight allocated to the fortresses. The number of aircraft available to the army was 244, the navy having only 20 or so. However, according to some estimates the Air Service had lost some 140 planes by the autumn of 1914, the inventory of the SW Front alone being reduced from 99 to 8 aircraft. By the end of 1914 Russia had established the so-called Air-Ships Squadron to operate the four-engined heavy bomber known as the Illya Muromets.

Although the air war over the Eastern Front was active and fierce it was a somewhat one-sided affair, dominated by the Central Powers. Russian production and repair facilities were limited; and between 1915 and the end of 1917 Russia imported, mainly from France, 1,800 aircraft and 4,000 engines. To ease production difficulties

(continued opposite)

the Russians resorted to the expedient of using captured enemy aircraft; at one stage the XVIII Corps Aviation Company was operating exclusively with captured machines. On 9 December 1917 the Air Service had 579 operational aircraft.

OPPOSITE A typical imported scout type, a Morane 'Parasol' monoplane. Russia's leading ace, Staff Capt.A.A.Kazakov (17 confirmed victories, but possibly as many as 32), flew an MS5 as commander of the XIX Corps air squadron, and later a Nieuport 17 as CO of the 1st Fighter Group of four squadrons.

The pilot officer is dressed in the flying kit issued from 1913: cf Plate F4. The helmet is brown leather with the usual officer's pattern cockade, the leather jacket is black, and the breeches are black piped red, indicating the Air Service's origin as a part of the Engineers. A double-headed eagle of the Engineer pattern with a bronzed metal twin-bladed propeller was the shoulder strap badge for pilots; it was worn in gilt metal by observers. Other ranks had the same badge stencilled in brown on their shoulder straps.

personal bodyguard of the Tsar, the *Konvoi*, declared their allegiance to the new regime within days of the abdication, symbolically removing the Imperial monogram from their shoulder straps.

Grenadier platoons

By the end of 1915 trench warfare led to the development of units which became known as 'grenadiers'; but note that these had no connection with the traditional Grenadier Corps itself. The XXV Army Corps raised the original unit in late 1915. Such detachments were to be a fourth platoon in each company, consisting of 'brave and energetic men' armed with ten hand grenades, a shovel and wire-cutters. Other detachments were armed with revolvers, carbines, swords or short lances. The purpose of the grenadiers was to lead intelligence-gathering raids, assaults and counter-attacks. During assaults they were to be found operating with the sappers, infiltrating and expanding the gaps in the wire. The extent to which the grenadier platoon system was instituted throughout the army is difficult to gauge, but they were certainly established in the Special Army, the Grenadier Corps and the XXV Army Corps.

St George's Battalions

Although not front line units, these hand-picked battalions were regarded as elite troops in that all ranks had to have been highly decorated: enlisted men with the St George's Cross or Medal, officers with the Order of St George. Formed as part of the guard at STAVKA in 1916, the original battalion was increased to five during July 1917, stationed at Minsk, Kiev, Pskov and Odessa and at STAVKA. Their task evolved into that of instructors to the Storm Battalions and other volunteer units.

Their uniform was the standard field dress with distinctions in the colours of the Order of St George, orange and black, based on the uniform of the 13th Dragoons (itself named the 'Military Order Regiment'). Officers had orange-piped breast pockets, breeches and cuffs; other ranks had orange-piped cuffs and breeches and also the front edge of the tunic. The officer's cockade had the officers' St George's Cross superimposed, and other ranks' their cross.

'Storm' and 'Death' Battalions

In the wake of the March Revolution the armed forces became a hotbed of political discussion, active warfare taking a back seat. However, this did not apply to all; and by May 1917 several proposals had been put to STAVKA to prevent the further deterioration of the forces. The impetus behind this movement came from the lower echelons and their suggestions were not always greeted with enthusiasm by higher command. However, it was decided to harness this enthusiasm, and various units were recruited during the build-up to the 1917 summer offensive.

There were two identifiable sources of such volunteers: troops already serving in combat units, and men either not in uniform or posted in rear areas. The second group was inspired to harness the revolutionary fervour of the population in support of what Kerenski called 'the freest army in the world'. Recruiting was carried out by members of the fabulously titled 'Executive Committee for the Formation of Revolutionary Battalions from the Rear', and enjoyed the support of some leading generals – notably, Brusilov. During the next six months 39 such battalions were formed. Some, such as those organised by cadet schools or combat formations (e.g. the 2nd Orenburg from Siberia), performed with great valour. Broadly speaking they were called 'Storm Battalions', 'Shock Battalions' or 'Battalions of Death'. Their purpose was to group together volunteers who were prepared to attack, and so inspire their comrades to follow.

The 1st Storm or Shock Detachment was formed on the South-Western Front commanded by Gen.L.G.Kornilov. The unit consisted of two 1,000-man battalions with three machine gun teams of eight guns apiece, and one foot and one mounted scout detachment of 16 men each. It performed well during the summer offensive but sustained very heavy casualties. When Kornilov became Supreme Commander-in-Chief one of his first actions was to reorganise the 1st Shock Detachment into the Kornilov Storm Regiment of four battalions (see Plate H2). The combat record of the regiment was such that for an action on 16 August every man in its ranks was recommended for the St George's Cross. Following the failure of Kornilov's so-called *putsch* the regiment was renamed the 1st Russian Shock, and later the Slav Shock Detachment, the latter possibly in honour of the number of Czech members of the unit.

Women's 'Battalions of Death' were also raised, but they were mainly a propaganda exercise. However, the remains of the 1st Women's Battalion (see plate H4) provided the Provisional Government's last line of defence outside the Winter Palace in November 1917 on the night of the Bolshevik coup. Naval 'Battalions of Death' (including one of women) were also raised, but details are scant.

The 'Savage Division'

Following the success of native volunteer cavalry units during the Russo-Japanese War it was decided in August 1914 to recruit a six-regiment cavalry division from amongst the Moslem tribesmen of the Caucasus and the Tartars of Baku. The official title was the Caucasian Native Cavalry Division, but it was also known as the 'Wild' or 'Savage Division' – both for its ferocity in battle, and also because many Russians regarded Caucasians in that light. The Tsar's brother, the Grand Duke Mikhail, commanded the division from 1914 to 1916. The organisation was only changed by the addition of a battalion of Ossetian rifle-men early in 1917.

1st Brigade Daghestan Regt, Kabardian Regt
2nd Brigade Chechen Regt, Tartar Regt
3rd Brigade Circassian Regt, Ingush Regt

A *divizion* of mountain horse artillery (two 6-gun batteries) provided support until 1916, when three

The founder and commander of the 1st Russian Women's Battalion of Death, Lt.Maria Botchkareva, reviewing the battalion in July 1917. Raised with the intention of shaming the male soldiers into adopting a more aggressive attitude to the war following the take-over by the Provisional Government, this battalion took part in the Kerenski Offensive of July 1917; see Plate H4. Similar women's units were formed, but only this first, from Petrograd, saw action.

batteries of Don Cossack horse artillery were added. The machine gun detachment was drawn from sailors of the Black Sea Fleet, who wore the kaftan and fur or fleece cap but with naval badges of rank. The signallers, transferred from the engineers, wore the standard army field dress. The officers had transferred from other cavalry regiments and were allowed to wear the cap of their original regiment when off-duty.

The 'Savages' fought on the South-Western and Romanian Fronts until transferred to take part in the Kerenski Offensive of summer 1917 as part of the 3rd Cavalry Corps. After participating in the abortive Kornilov coup the division did not return to the front, and was finally sent home to demobilise later that year.

NON-RUSSIAN UNITS

Belgian

The stagnation on the Western Front made the armoured car units of both Britain and Belgium surplus to requirements. During summer 1915 the Russians requested that the Belgians transfer a unit to the Eastern Front, where there was a pressing need for such weapons. By December 1915 the *Corps des Autos-Canons-Mitrailleuses Russie* were parading for Nicholas II at Tsarskoe Selo. The vehicles were organised into a battery of two sections, each of five armoured cars – three Mors and two Peugeots – armed with 8mm machine guns and 37mm short guns. The men of all ranks were dressed in a mixture of Belgian and Russian uniform with rank distinctions worn on the shoulder straps. The other ranks wore volunteer cords on their shoulder straps, which entitled them to certain privileges. Contemporary photographs show that the Belgian stable-cap was commonly worn. The *Corps* fought on the South-Western Front under 9th Army until summer 1917. The majority of the 360 men returned to Belgium via Vladivostock during summer 1918.

British

Shortly after the arrival of the Belgians a British armoured car unit landed at Alexandrovsk (Murmansk) on New Year's Day 1916. Known as the Russian Armoured Car Division, Royal Naval Air Service, it was led by Commander Locker-Lampson MP. During the summer and autumn of 1916 it served on the Caucasian Front fighting the Turks, and gained valuable experience before being transferred to the South-Western Front in Romania, where it went into action in December. Relocated in June 1917 as part of the build-up for the Russian summer offensive, the RNAS unit participated in both the advance and the retreat. The vehicles were mostly machine gun-armed Lanchesters, supplemented by Pierce-Arrow armoured lorries mounting 3-pounder guns. Following the debacle of the Kerenski Offensive the British vehicles were supplemented with Fiat armoured cars and lorries. In autumn 1917 the majority of the men were withdrawn via Kursk and Archangel to Britain.

During 1917 a mission was sent by the Royal Flying Corps to train Russian air and ground crews in the use of British aircraft supplied to Russia. Allocated to the South-Western Front in time for the Kerenski Offensive, they were withdrawn in December 1917. (A French aviation training mission was also operating in the same area at the time.)

The Navy was a hotbed of political activity following the abdication of the Tsar in spring 1917, in part due to their secondary role during the war. The Baltic Fleet, based at Kronstadt, was largely limited to coastal operations under command of Northern Front. Its sailors were among the most fervent supporters of the Bolsheviks in November 1917 (and their most courageous opponents in 1921). The Black Sea Fleet, based at Sebastopol, saw slightly more wide-ranging action against the Turks. Combined operations in April 1916 at Trabzon on the coast of Anatolia were successful, but the scope for more ambitious operations was limited.

This member of the crew of the *Diana*, sister ship of the cruiser *Aurora* of revolutionary fame, is dressed in the traditional naval manner. Of particular interest is the black and white *telniashka* vest worn under the blue jumper; when stripped to this vest in battle, the man could not retreat or surrender.

A number of privately funded, voluntary medical units from Britain were active with the Russian army until the end of 1917.

* * *

The units formed from Russia's subject races rarely wore anything to distinguish them from other Imperial troops. The grudging manner in which they were raised typified the Imperial government's distrust of anything that hinted at nationalist aspirations. The majority of these formations were no more than tokens until the March Revolution ushered in a more liberal atmosphere. Even units formed from prisoners taken from the Austro-Hungarian army were treated in the same manner.

Polish

Russia ruled the greater part of Poland, and in October 1914 unenthusiastic permission was given to raise a Polish Legion for service with the Russian army. However, it was relegated to the status of a reserve infantry battalion and two squadrons of lancers. Due to the manpower shortage it was agreed, in January 1917, to expand these units to a Rifle division and a lancer regiment. During the summer offensive of 1917 they were held in reserve and some elements were used to control the retreat that followed. The Provisional Government further upgraded the Poles to an army corps, which became I Polish Corps in September 1917. The corps was based in Byelorussia when the treaty of Brest-Litovsk was signed in March 1918. The majority of the Polish troops surrendered to the Germans, though some fought with the anti-Soviet forces in southern Russia during 1918.

Latvian

As part of Russia's Baltic territories Latvia was directly in the line of Germany's advance up the Baltic coast in 1915. On 1 August 1915 nine Latvian Rifle battalions were raised, with Latvian officers. During 1916 the battalions were grouped first into a single brigade and then into two, each of two regiments. By the beginning of 1917 the brigades were consolidated into a division as part of VI Siberian Corps on the Northern Front. Throughout their service in the Russian army the Latvians fought on their home territory. However, the division fell under the influence of Bolshevik agitators and played a significant role during the Civil War as part of the Red Army.

Serbian

During late 1915 and early 1916 permission was granted to form a Serbian infantry division from among the Slavs captured from the Austro-Hungarian army. The division was assembled near Odessa and numbered some 10,000 men and officers.

This lieutenant of the Air Service is an artillery observer, as shown by the crossed cannons on his shoulder straps. He is wearing the 1916 woollen shirt-tunic with the Officer's Cross of St George and the Cross of St Vladimir 4th Class. The St George's Cross was Russia's highest award for valour. It came in two categories – one for officers, one for other ranks – each of four classes.

Czech

Probably the most famous of the 'foreign' units, the Czech Legion was raised in August 1914. Initially consisting of four companies, the unit was committed to the Galician theatre. The Tsarist government did not encourage its expansion until late 1915, when it became the Czechoslovak Rifle Regiment with eight rifle companies and a trench mortar company, some 1,600 men in total. Following the March Revolution the Provisional Government authorised the expansion of the regiment to divisional strength by recruiting from amongst Austro-Hungarian prisoners of war. By the time it went into action during the Kerenski Offensive the 'Czechoslovak Army Corps' numbered 7,000.

UNIFORMS & PERSONAL EQUIPMENT

Until the early years of the 20th century boots and clothing had been made by the troops themselves from materials provided by the government. This 'regimental economy' led to the waste of much time that should have been used in training; and poor quality control had left many soldiers convinced that the government was indifferent towards them. Following the Russo-Japanese War it was decided to phase this system out. By 1909 some 50 per cent of production had been taken over by contractors theoretically supervised by the Quartermaster's Department. Officers provided for themselves from military outfitters in the large cities, and the quality of their uniforms was consequently much better than that of the men's.

Cossacks were also expected to provide their own uniforms, as were the other irregular cavalry formations. The service dress of Caucasian Cossacks was based on their traditional clothing, more native Caucasian than Russian in style (see Plate E3).

Clothing the Imperial Militia was the responsibility of the local government in the relevant district. The only regulations were that the men had to be uniformly dressed; and that scarlet shoulder straps and the 'militia cross' on the service cap (*furashka*) were compulsory. Consequently, certainly during the early months of the war, many men were dressed in obsolete clothing such as the white shirt-tunic and dark green breeches (*sharovari*).

Parade uniforms and other orders of dress would occupy volumes in themselves, therefore any reference to them will be limited to those items that were worn by the combat troops – e.g., it was common for officers, and not only in 1914, to wear their parade caps in the field.

Service dress

In 1907 khaki of a light olive-green shade was introduced as the service dress colour for all ranks

This posed group portrait of three privates of the 67th Infantry Regt is interesting in showing three variants of the shirt-tunic. From left to right they are the M1910 in wool, the M1912 in wool and the M1914 in cotton. Both the buttons and the position of the fastening vary from style to style. The private in the centre is wearing a peakless cap with a chinstrap, indicating a mounted role of some sort. The breeches were cut to fit snugly into the boots.

Siberian infantry regiments were all termed Rifles. The private shown here is dressed for cold weather in the *shinel* greyish-brown greatcoat with the *bashlik* cowl neatly arranged across his chest. The headgear is the Siberian fleece hat, much 'woollier' than the *papaha*. Collar patches, cowl piping and shoulder strap piping all show well in this excellent portrait.

Table A: Guards tunic piping & lace

All ranks wore cuff piping or lace; officers' pocket flaps were piped in the same colours as the lace down the opening front edge of other ranks' shirt-tunics.

1st Division: cuffs – white
Pocket & front: 1st Regt – scarlet; 2nd – light blue;
                  3rd – white; 4th – dark green
2nd Division: cuffs – red
Pocket & front: 1st Regt – scarlet; 2nd – light blue;
                  3rd – white; 4th – dark green
3rd Division: cuffs – yellow
Pocket & front: 1st Regt – yellow; 2nd – light blue;
                  3rd – white; 4th – dark green
Guards Rifles:
Cuffs: 1st Regt – crimson; 2nd – white
Pocket & front: 1st to 4th Regts – crimson

Guard Foot Artillery:
Cuffs only: 1st Bde – white; 2nd – blue; 3rd – yellow
Guard Horse Arty: all piping black
(Among Guard Cossack units only the Combined Cossack Regt had cuff lace, in yellow.)

Table B: Cavalry breeches stripes

Guard cavalry: Scarlet, except:
Hussars – white
His Majesty's Cuirassiers – yellow
Her Majesty's Cuirassiers – light blue

Line cavalry:
Dragoons
Scarlet   1st, 3rd, 4th, 9th, 10th, 13th, 15th, 16th, 17th,
              18th, 19th Regts
Light blue   6th, 11th, 12th Regts
Yellow   5th, 7th, 8th, 20th Regts
Light green   14th Regt
Pink   2nd Regt

Lancers
Scarlet   1st, 5th, 9th, 15th, 16th, 17th Regts
Light blue   2nd, 6th, 12th, 14th Regts
Yellow   4th, 8th, 10th, 13th Regts
White   3rd, 7th, 11th Regts

Hussars
Yellow   1st, 2nd, 3rd, 4th, 9th, 10th, 11th, 12th, 16th,
              17th Regts
White   5th, 6th, 7th, 8th, 13th, 14th, 15th, 18th Regts

Daghestan Cavalry Regt – blue breeches with scarlet stripe
Ossetian Divizion – plain
Turcoman Division – yellow stripe

and branches of the regular army. After washing and hard wear the khaki would fade – almost to white, in the heat of summer on the South-Western and Caucasian Fronts.

Officers' service dress consisted of a tunic, *kittel*; breeches, knee-high boots and peaked cap. The khaki tunic was single-breasted, in cotton for summer and wool for winter, with five buttons either of metal or leather. It had two inside skirt and two outside breast pockets, the flaps of the latter cut to a point *en accolade*. The tunic had a stand collar 45mm high, fastened with two hooks; the cuffs were round for foot and curved to a point for mounted branches. Guards infantry officers' breast pocket flaps were piped to indicate the number of the regiment within the division, and cuff piping identified the division – see **Table A**, this page.

Officers' shoulder boards or *pogoni* – such a potent symbol of their privileged status that the Revolutionary regime banned them in 1918 – were at first stiff, detachable items, faced with bright metallic lace. The ranks were distinguished by a system of metallic stars and lengthways lines of the base colour, and the unit and branch of service by additional insignia – see **Tables F, G & H,** pages 33 & 35.

The breeches were cut to fit comfortably inside the boots; they were dark 'Tsar's green' for infantry and other dismounted troops. Cavalry officers wore blue-grey, or sometimes regimentally coloured breeches – crimson for hussars and dark blue for all others. Cavalry breeches were striped in the regimental colours – see **Table B**, page 22. Unstriped khaki breeches were clearly more practical in the field and became almost universally worn as the war dragged on. Mountain artillerymen wore plain black leather breeches.

Each Steppe Cossack Host wore a distinguishing colour; this was particularly noticeable as a broad stripe on their dark blue breeches. See **Table C**, this page.

Other ranks had been issued a soldier's version of the officer's tunic, without breast pockets, until 1912. Officially discontinued, it was, however, seen in use throughout the war. The almost universal wartime garment was a pull-over shirt-tunic, the *gymnastiorka*, based on the smock-like shirt of the Russian peasant and worn untucked over the breeches, with knee-high boots and peaked cap.

There were several versions of the shirt-tunic, produced both by outside contractors and the regimental economies. Usually such variations took the form of the addition of one or two breast pockets, and the position of the front fastening – either vertically from the centre of the collar or off-set to either side. There were normally five small front buttons of horn or metal. Summer and winter shirt-tunics were produced in cotton or wool respectively. Cuffs were either plain, or shirt-fashion with two buttons. The only line infantry who wore cuff lace were the machine gunners and the scouts, in crimson and green respectively. Guards infantry wore cuff lace to identify their division, and lace down the front opening to identify the regiment within the division – the colours followed those of their officers' cuff and pocket flap piping respectively, as in **Table A**.

Shoulder straps indicated rank, branch of service, unit number and other information. The shoulder straps were reversible, one side khaki and the other appropriately coloured. They were worn on both greatcoat and tunic. See **Tables E & F**, pages 24 & 33, for basic details.

In bad weather all ranks and most branches wore the greatcoat, *shinel*; a cap of natural fleece or artificial astrakhan lambswool, *papaha*; and a shawl-like cowl, *bashlyk*. In addition the Cossacks and other irregular cavalry of the Caucasus wore a sleeveless black cloak of felted goat or camel hair, the *burka*.

A very young member of the 23rd Engineer Bn; the number and the arm-of-service badge can be seen on the shoulder straps. His gymnastiorka is the cotton M1914 pattern.

Table C: Cossack distinctions

Host	trouser stripe	shoulder strap/ piping	collar patch/ piping
Don	scarlet	dk.blue/ scarlet	scarlet
Orenburg	light blue	light blue	light blue
Ural	crimson	crimson	crimson
Siberian	scarlet	scarlet	scarlet
Astrakhan	yellow	yellow	yellow
Amur	yellow	green/ yellow	green/ yellow
Semirechensk	crimson	crimson	crimson
Ussuri	yellow	yellow/ green	yellow
Trans-Baikal	yellow	yellow	yellow
Foot Artillery	–	scarlet	black/ scarlet
Horse Arty.	–	dark blue	dark blue?
Kuban*	scarlet	scarlet	scarlet
Terek*	light blue	light blue	light blue
Kuban inf.		crimson	black/ crimson
(* Inch wide stripes only)			

Each regiment had an individual monogram worn on the shoulder straps; to list them all here would be impossible, but some examples (of the Latin letters which resemble the Cyrillic initials) are:
Kb = Kuban, O = Orenburg, y = Ussurski, A = Amur,
Cm = Semirechi, 3b = Trans-Baikal, Br = Terek. The Don units displayed a Cyrillic D, to distinguish them from Dragoons, which used the Latin initial.

The greatcoat was made of blue-grey cloth for officers and coarse grey-brown wool for other ranks. It was double-breasted, with a fall collar, fastening on the right with hooks-and-eyes. An earlier model had a single row of six metal buttons down the front; although manufacture was discontinued before the war they were worn as long as stocks lasted. The coat was generously cut, and gathered at the back by a half-belt and two buttons. For dismounted men the greatcoat reached halfway between knee and ankle, with a loosely sewn turn-back hem which could be turned down in extreme weather. Mounted troops wore a longer type, with cuffs curving to a point at front and back; one of these was traditionally worn unsewn, to carry messages. Coloured collar patches were applied to the greatcoat; in some units these were edged with coloured piping to identify the regiment and branch of service – see **Table D**, this page. Officers' and NCOs' patches bore a button in regimental metal.

The peaked (visored) cap was available in both coloured and service versions. The service type was khaki with a black peak – this was painted green in the field. Officers and mounted personnel had chin straps, others did not.

The basic shade of the coloured version for dismounted units was dark green. In the Guards the band was coloured in regimental sequence within the division – red, blue, white and green for the 1st to 4th Regiments. The Grenadiers and Line infantry used the same system. The band and crown seam were piped in red. The artillery and technical branches had a black band, and scarlet piping around the band

Table D: Greatcoat collar patches

Unit	patch	edging
1st Guards Inf Div:		
1st Regt	scarlet	–
2nd Regt	blue	scarlet
3rd Regt	white	scarlet
4th Regt	green	scarlet
2nd Guards Inf Div:		
1st Regt	scarlet	green
2nd Regt	blue	–
3rd Regt	white	–
4th Regt	green	–
3rd Guards Inf Div:		
1st Regt	yellow	–
2nd Regt	blue	yellow
3rd Regt	white	yellow
4th Regt	green	yellow
Guard Rifles:		
1st Regt	crimson	–
2nd Regt	green	crimson
3rd Regt	green	crimson
4th Regt	green	crimson
Guard Cavalry	(regimental colours)	
Guard Cossacks	(regimental colours)	
Guard Artillery	black	scarlet
Guard Engineers	black	scarlet
1st–4th Grenadier Divs, & all Line Inf Divs:		
1st Regt	scarlet	–
2nd Regt	blue	–
3rd Regt	white	–
4th Regt	green	–
Line Rifles	green	crimson
Line Cavalry	(regimental colours)	
Line Artillery	black	scarlet
Line Engineers	black	scarlet

Table E: Other ranks' shoulder strap colours

Unit	coloured side	piping, khaki side
Guard Infantry	scarlet	regimental (Table A)
Guard Rifles	crimson	
Guard Cavalry	as breeches stripe, regimental piping	
Guard Cossacks	regimental colours	
Guard Arty.	scarlet	–
Guard Engr.	scarlet	–
1st Gren Div	yellow	–
2nd Gren Div	yellow	–
3rd Gren Div	yellow	–
4th Gren Div	yellow	–
Line Inf Divs:		
1st Bde	scarlet	–
2nd Bde	blue	–
Line Rifles	crimson	–
Line Cavalry	as breeches stripe, piped:	
Dragoons		dark green or white
Lancers		dark blue
Hussars		–
Line Arty.	scarlet	–
Line Engr.	scarlet	–
Gren.Arty.	scarlet	–
Gren.Engr.	scarlet	–
Cossacks	(Table C)	
Daghestan Cav	scarlet	–
Turcoman Cav	yellow	–

STAFF
1: Captain, General Staff, 1917
2: General of Artillery Irmanov, 1914
3: Adjutant General of Cavalry Brusilov, 1916
4: Colonel, medical service, 1915–16

A

INFANTRY
1: Lieutenant-Colonel, 94th Yeniseiski Regt, 1914
2: Private, 404th Kamyshinskiy Regt (Opolchenie), 1915
3: Senior NCO, machine gun *kommando*, 8th Moscow Grenadier Regt, 1917
4: Senior private, grenadier platoon,
   4th Rifle Bde ('Iron Brigade'), 1916

LINE CAVALRY
1: Bombardier Layer, 20th Horse Artillery, 1915
2: Captain, 5th Alexandriyski Hussar Regt
   ('Immortal Hussars'), 1916–17
3: Trooper, 16th Tverskoi Dragoon Regt, 1915–17

COSSACKS
1: Cossack, 1st Argun Regt, Trans-Baikal Host
2: Cossack, Kuban Cossack infantry
3: Lieutenant, 2nd Volgski Regiment, Terek Host
4: Warrant Officer, 17th Don Cossack Regt General Baklanov, 1914

SPECIAL TROOPS
1: Armoured car driver, 7th Automobile Machine Gun Platoon, 1915
2: Stretcher bearer, 1915–17
3: Regimental Orthodox priest
4: Pilot officer, Aviation Service, 1914
5: Cyclist, 3rd Bicycle Company, 1915–17

NATIONAL TROOPS
1: Private, 5th Latvian Rifle Regt, 1916–17
2: Trooper, Turkmen Horse Half-Regt, 1914–15
3: Trooper, 'Savage Division', 1914–17
4: Lieutenant, Polish Lancers, 1917

ELITE UNITS 1917–18
1: Lieutenant, Shock Bn of Rear Echelon Volunteers, 1917
2: NCO, 1st (Kornilov's) Shock Regt, 1917
3: Russian 'Legion of Honour'; France, 1917–18
4: 1st Women's Death Bn, 1917
5: 2nd Vol.Det. of Crippled Warriors, 1917
1A: Shock Bn of Rear Echelon Volunteers
5A: Committee for Recruiting Disabled Soldiers
6: Reval Naval Shock Bn
7: St George's Bn, STAVKA
8: 1st St George's Bn, Kiev
9: Death Bns
10: Shock Bns

and crown. The permutations of the coloured cap in the cavalry branch were virtually endless, based on individual regimental colours; for instance, in Line lancer regiments the base colour was blue with the band and crown piping as the trouser stripe.

Pressed metal cockades were worn at the front centre of the cap band. There were three qualities, for officers, NCOs and enlisted men; the colours were the Romanov orange, black and white. Militia units wore the 'Opolchenie cross' above the cockade. The cockade was also worn on fleece caps which were the standard winter headgear for the army. Siberian units and other groups wore variations on it – sometimes larger, darker and shaggier, or smaller like the *kubanka* of the Kuban Cossacks, which came into use due to a shortage of material.

Personal equipment

In 1912 the officer's field kit was designed around a brown Sam Browne-style belt, with two braces worn vertically at the front and crossed at the back. The sword hung from the left hip in the Oriental manner. On the left strap was a whistle, on the right side a pistol holster. Map cases and binoculars, often privately purchased, completed the outfit. When mounted the greatcoat was strapped to the front of the saddle. Haversacks were usually carried with the baggage.

Other ranks' equipment consisted of a leather waist belt, white for the Guards and brown for all other troops, supporting on each side of the belt plate 30-round ammunition pouches, and on the right side the Linnemann entrenching tool slung with the handle down. Suspended over the right shoulder was a waterproof

Table F: Shoulder strap devices

Officers had metal or embroidered bullion devices, for foot units in the same metal as buttons and lace, for mounted units in the opposite metal. **Other ranks** wore the same devices stencilled in coloured paint. The colours listed below appeared on the khaki side of the straps; infantry wore yellow stencilling on the coloured side, unless this was white, yellow or dark blue, when the stencilling was scarlet.

The straps bore a range of devices: at the outer end, unit numbers, often with Cyrillic initial letters of categories of troops, or where appropriate, regimental monograms referring to the colonels-in-chief; and nearer the neck, for specialist branches, branch-of-service badges. Sometimes combinations were worn: e.g., the 5th Rifle Artillery wore crossed scarlet cannons above crimson '5.C.' – the C being the Cyrillic S for *Streltsi*, Rifles.

There were many permutations of category initial letters; some examples, again using the Latin letters most resembling the actual Cyrillic forms, were: T = hussars, y = lancers, Latin D = dragoons (to avoid confusion with the Don Cossacks, who used the Cyrillic form), K = Caucasian, T = Turcoman, Cb = Siberian. Militia units wore M preceded by the letter of the district, but identification was problematic since so many districts began with the same initials.

Branch	colour	device
Corps staff	orange	Corps no., Roman numerals
Line inf.	yellow	unit no. or monogram
Cyclists	yellow	unit no., & badge of crossed rifles on bicycle
Machine guns	yellow	unit no., & machine gun
Gren.inf.	scarlet	monogram only
Gren.arty.	scarlet	grenade on crossed cannons, unit no.
Rifles	crimson	unit no. or monogram
Rifle Arty.	crimson	(as Artillery)
Cavalry	lt.blue	category initial, unit no. or monogram
Cossacks	dk.blue	*Voisko* initial, unit no. or monogram (see Table C)
Artillery	scarlet	crossed cannons, bde.no. in Roman numerals, monogram if appropriate
Fortress arty.	orange	crossed cannons above fortress initial
Horse arty.	lt.blue	crossed cannons, unit no.
Horse mtn.arty.	lt.blue	(as Horse Artillery)
Engineers	brown	crossed pick & shovel, unit no. or monogram
Gren.engrs.	brown	grenade on crossed pick & shovel, unit no.
Pontoneers	brown	crossed pick, axe, saw & shovel on anchor
Odessa Naval Bn	green	crossed axe, shovel & anchor
Armoured cars:		
1914–15	green	machine gun over winged wheels, steering wheel, unit no.
1916–17	scarlet	
Aviation	brown	winged propeller
Signals*	brown	two entwined lightning bolts
Signals (radio)	brown	winged *foudre* on entwined lightning bolts
Railway troops	green	crossed axe & anchor, unit no.
Supply & Trspt (inc.hospital trains)	white	–?–
Field hospitals	orange	Cyrillic L above unit no.
Intendance	black	Cyrillic I

* Artillery signallers showed their specialism by a cloth badge of crossed signal flags in red worn above the left elbow; artillery telephonists – later all telephonists – by a similar badge of entwined lightning bolts.

canvas haversack (replaced by a knapsack slung behind the shoulders in the Guards), which contained clothes, food and other personal items. The greatcoat was rolled and carried horseshoe-fashion over the left shoulder, with a spare pair of boots and the cowl rolled up in it. An aluminium water bottle and the oval mess tin were suspended over the right shoulder, though the latter is often seen with the greatcoat ends tucked into it. Each man carried one-sixth part of a tent and its poles attached to the coat roll. The weight of all kit including ammunition was some 56½ pounds.

Wartime innovations
The functional design of the field service dress in use in 1914 meant that very little alteration was necessary during the war; the changes introduced were not nearly as sweeping as those in other European armies. A system of wound stripes was introduced in 1916, and some other insignia such as those marking the 'grenadier platoons' (see Plate C4).

Officers in the field made themselves less conspicuous by modifying their shoulder straps. During the war a fashion developed replacing the stiff detachable shoulder boards with soft straps that were sewn into the shoulder seams. The conspicuous metallic lace was replaced by a subdued khaki equivalent; or rank insignia were even drawn onto plain cloth straps with indelible pencil. It became common for officers in the field to wear the soldier's shirt-tunic and to discard their swords; another fad was to remove the stiffener from the cap to give it a softer, more British look. During the war years many officers also purchased tunics which, while retaining the stand collar, were otherwise modelled on the British officer's service dress jacket: generously cut, with two large, pleated patch pockets on the breast and two very large 'bellows'-style expandable pockets on the skirt. This was confusingly known as the 'French', in reference to Gen. Sir John French, who commanded the British Expeditionary Force in France in 1914. General Staff officers also began to wear a special black tunic which, though apparently unofficial, existed in several versions.

At the end of 1916 the army adopted the *pilotka*, based on the forage cap of the Aviation Service, to be worn under steel helmets. From the spring of 1917 they were issued to all officer cadets. Steel helmets of the French Adrian pattern, with an Imperial eagle badge added to the front, were imported from 1916 until domestic production facilities could be established. Although they were available, steel helmets were not popular and generally seem to have been worn mainly by grenadier units and 'death battalion' personnel during 1917.

A shortage of leather necessitated the introduction of ankle boots worn with puttees. The only items of equipment introduced during the war were a gas mask, a fabric ammunition bag worn suspended from the shoulder, and a 60-round canvas bandolier.

Following the Revolution of March 1917 men of all ranks, depending on their political persuasion, removed all Tsarist symbols from their uniforms – particularly the regimental monograms on the shoulder straps associated with members of the imperial family. The Provisional Government introduced an alternative system of rank indicators for officers similar to the cuff rings of the navy; however, these do not seem to have enjoyed widespread acceptance (see Plate H1).

One of the variations of cotton shirt-tunic, in this case the M1916. The peak of the khaki service cap has been painted green to cut down on reflection, and the shoulder straps are worn khaki side up. The M1904 belt had a metal other ranks' buckle plate with the Imperial double-headed eagle for the infantry, an Imperial eagle with crossed cannon for the artillery, crossed axes for the engineers, and a flaming grenade for the grenadiers. Guard regiments had their own distinctive pattern.

Flags

Each infantry regiment had a colour, which was carried to the front by the 1st Battalion. Regiments also had a camp colour, measuring 50ins. x 35ins., coloured according to the regiment's place in the divisional sequence and bearing its number in black in the centre. Battalions and companies also had flags, which were carried on the bayonets of the men acting as markers. Those of battalions had three horizontal stripes of black, orange and white with the battalion number on the central stripe. Company flags were coloured according to the number of the regiment within the division, and had a vertical and a horizontal stripe which crossed centrally. The horizontal stripes were red, blue, white and dark green for the 1st–4th battalions respectively; the vertical stripe, in the same colour sequence, identified the 1st–4th companies in each battalion. (For instance, a white flag with a cross formed by a scarlet vertical and a blue horizontal stripe was that of the 1st Co, 2nd Bn, 3rd Regiment in its division.)

All Line Cossack regiments carried two banneroles into action, one to mark the regimental commander, the other the squadron. The regimental marker was a 35in. square in the same colour as the shoulder straps. Those of the Siberian, Orenburg, Semiretchi and Trans-Baikal regiments bore a

Table G: Officers' shoulder strap lace colours

Unit	Gold or Silver
1st Guards Inf Div	G
2nd Guards Inf Div	G
3rd Guards Inf Div	S
Guard Rifles:	
1st & 3rd Regts	S
2nd & 4th Regts	G
Guard Cavalry	(regimental)
Guard Foot Artillery	G
Guard Horse Artillery	G
Guard Engineers	S
Guard Cossacks	(regimental)
1st, 2nd, 3rd Gren Divs	G
4th Gren Div	S
Line Infantry	G
Line Rifles	G
Line Cavalry	(regimental)
Line Artillery	G
Line Engineers	S

Table H: Rank insignia

Rank was shown on the shoulder straps at all levels. (The Cossacks, and in some cases the cavalry, used the alternative titles shown in the notes below, but the same insignia.) **Other ranks** wore strips of lace attached across the top of the shoulder straps; pre-war these were in either yellow or white, and orange for the Guards, but at the beginning of the war red replaced the other colours. **Officers'** shoulder boards had narrow lengthways stripes of base colour showing between the strips of metallic lace facing – central single stripes, and double stripes dividing the width into thirds – and added five-point metal stars. **General officers** had metallic lace facing in a zig-zag pattern, without lengthways stripes of base colour. It was a Russian peculiarity that the senior, rather than the junior rank within a grade wore no stars.

Russian rank	British equivalent	insignia
Ryadovoi (1)	private	–
Yefreitor (2)	lance-corpl.	1 stripe
Mladshi Unteroficier (3)	corporal	2 stripes
Starshi Unteroficier	sergeant	3 stripes
Feldfebel (4)	sgt.major	1 wide transverse metallic stripe
Pod-Praporshchik (5)	warrant officer	1 wide lengthways metallic stripe
Pod-Poruchik (6)	2nd lieutenant	2 stars, 1 stripe
Poruchik (7)	lieutenant	3 stars, 1 stripe
Shtabs-Kapitan (8)	staff captain	4 stars, 1 stripe
Kapitan (9)	captain	1 stripe
Maijor	major	2 stars, 2 stripes
Pod-Polkovnik (10)	lt.col.	3 stars, 2 stripes
Polkovnik	colonel	2 stripes
General-Maior	maj.gen.	2 stars
General-Leitnent	lt.gen.	3 stars
General	general	–
General-Feldmarshal	field-marshal	crossed batons

Notes:
(1) Cossacks, *Kazak*. (2) Cossacks, *Prikazni*.
(3) Cossacks, *Uryadnik*. (4) Cossacks and cavalry, *Vakhmistr*.
(5) Cossacks, *Pod-Khorunji*. (6) Cossacks, *Khorunji*; cavalry, *Kornet*. (7) Cossacks, *Sotnik*.
(8) Cossacks, *Pod-Esaul*; cavalry, *Shtabs-Rotmistr*. (9) Cossacks, *Esau*; cavalry, *Rotmistr*.
(10) Cossacks, *Voiskovoi Starshina*.

St Andrew's cross in white, and the Amur Cossacks in yellow. All showed the regimental number in the centre.

The squadron markers were swallow-tailed, 22½ins. in the hoist, 35ins. in the fly to the tips of the tails and 15ins. in the notch. The upper half was in the regimental colour, the lower in that of the squadron, with a white or yellow median stripe if the regimental bannerole had a St Andrew's cross. Squadron colours were scarlet, light blue, white, dark green, yellow and brown for 1st–6th squadrons respectively.

TACTICS

Infantry

Offensive infantry tactics prevalent in most of the armies of Continental Europe at the outbreak of the First World War blissfully ignored the developments in weapons that had taken place during the previous 50 years. Men and their officers would line up, in close formation, and move at varying speeds across more or less open ground, to drive their foes before them with élan and the bayonet. Russia, with the recent experience of the Russo-Japanese War to draw upon, had theoretically instituted some tactical changes, but these were not in widespread use by the outbreak of war.

Before 1914 Gen.Lesh had developed a system based on platoons advancing in groups of three men with roughly two metres space between each man. Each trio moved independently so that no more than two groups were on their feet at once. The depth was three lines, with other platoons extending the line on either flank. By contrast, the usual method of attack in 1914 was the 'chain', with some two metres between each man and some six metres depth between each 'chain' – basically straight, extended lines. The bayonet charge began at 50 metres from the enemy line. Lesh's system was slow but efficient; the 'chain' was a recipe for heavy casualties. Machine guns were to be pushed well forward to support the advancing troops, and hand grenades were also to be used. However, the firepower available to defenders – and Russia's weakness in artillery support for much of the war – was such as to render all but the most carefully prepared attacks suicidal, as was so tragically the case across Europe during 1914–16.

The 1912 **cavalry** regulations stressed the importance of 'initiative and resolution', and that each trooper 'must be prepared to fight with his rifle in his hand as well as any infantryman'. However, the majority of cavalry officers still dreamed of knee-to-knee charges with sabre and trumpet. Cavalry-vs-cavalry actions would commence by trotting in extended order, breaking into a charge and closing ranks at between 100 and 50 metres from the enemy. The 'swarm', *lava*, was an old Cossack formation that had been adopted by the regulars to disrupt the enemy formation, break through a picket line, or

Gas warfare began on the Eastern Front at Bolimov during January 1915. Russia experimented with a variety of masks. The most commonly issued was perhaps the Koumant-Zelinski made from the second half of 1916, which is worn by the two right hand men here. The device consisted of a reddish rubber hood and a rectangular tin respirator canister supported by neck tapes; when not in use the hood was stowed inside one end of the canister, which was slung over the right shoulder. Other respirators were based on mine rescue equipment. The Russians used gas themselves, and by the end of 1916 had 15 chemical (gas) detachments.

to envelop an enemy force. A five-metre gap was left between the riders in the front rank, the second rank filled the gaps some 20 metres behind. It was against the Austro-Hungarians that the Russians fought their last great cavalry-vs-cavalry battle, including knee-to-knee charges, at Jaroslavice in Galicia on 20 August 1914.

A cavalry squadron attacking infantry or artillery would advance in single rank extended order. It is interesting to note that several successful attacks were carried out during the war against advancing infantry. As Gen.Danilov said, 'infantry naturally fire at the rider, not the horse'.

Dismounted action was also noted in the regulations, with one-third of the men being detailed as horseholders. It very rapidly became obvious that modern weapons made the old role of cavalry redundant. To dismount the men and turn them into infantry was one solution but, until it began to be adopted in late 1916, the cavalry was generally kept waiting in anticipation of exploiting a breakthrough that never really came.

The **artillery** was supposed to accompany infantry attacks, engage hostile artillery and machine guns, destroy obstacles and oppose counter-attacks. Conditions during the war demanded a different approach, and artillery turned into a fairly static sledgehammer to pound enemy defences.

Wartime innovations

The innovative use of the newly formed grenadier units to spearhead attacks was one of the ways in which the Russians sought to overcome the murderous obstacles to the success of frontal assaults. Another was the use of partisan warfare, particularly in the primeval conditions of the Pripyat Marshes; in this area vast swamps cut by numerous waterways prevented the Central Powers from establishing a continuous line. A raid carried out in the early winter of 1915 attracted attention: a force of irregulars, guided by sympathetic locals, launched a night raid on the HQ of a German infantry division and took prisoner the commander – who committed suicide – and several of his staff. This success led the *Ataman* of the Cossacks to order the formation of partisan units from volunteers in every Cossack regiment. The most famous of these was the 'Wolves', so called because their standard was the skin of a wolf with wolf-tails attached to the pole, and the men wore wolfskin caps. The unit was led by a young Kuban Cossack officer, A.G.Shkuro, who was later to become famous as a cavalry leader in the Civil War.

However, the major innovation was the method of assault used during the opening phases of the Brusilov Offensive in 1916.

These *frontoviki* at least have the benefit of dry weather. The fleece caps show that it is either autumn or winter. The man nearest to the camera is drawing five-round rifle ammunition clips, some of which will be kept in the waist pouches, some in the canvas bandolier across his chest – an item that became more common as the war dragged on. At his right hip can be seen a gas mask container. His entrenching tool is near at hand, and from the shine on its edge it is well sharpened in preparation for use as a hatchet in close combat. The private in the background has taken the precaution of covering the breech of his M1891 Mosin-Nagant rifle to protect the mechanism from mud. This sturdy, simple rifle was the standard infantry weapon of both the Imperial and the Red armies. As the shoulder straps of all the men in this photo are worn in the reverse (khaki upward) fashion it is impossible to tell their ranks or unit.

The Brusilov Offensive

General A.A.Brusilov was promoted to the command of the South-Western Front in April 1916, and immediately set about preparing an offensive against the Austro-Hungarians, who were well dug into positions which they had occupied for several months. Brusilov and his staff had analysed the failures of previous Russian offensives, however.

Artillery preparation would be co-ordinated so that the Austrians would be confused. Apparently random pauses and resumptions in the bombardment were calculated to throw them off balance and make them wary about emerging from their dugouts during a lull. The entire line would be bombarded in the same way so as to give no hint of where the main attack would fall, thus preventing the enemy from placing their reserves effectively for a counter-attack.

Trenches were sapped forward, in places to within 50 metres of the enemy, along as much of the line as possible. Huge dugouts were built to house the reserves within a short distance of the front line, to enable them to get forward quickly. Models were made of the Austrian defences in each sector, and the Russian infantry were trained in them so as to become familiar with their objectives. Careful aerial reconnaissance, including photography, was carried out, and enemy batteries were pin-pointed.

The South-Western Front's four armies faced a similar number of Austrians. Brusilov ordered each army to choose a point to attack depending on the local conditions. The offensive was to have begun in July, but was brought forward to relieve the Italians, who were under huge pressure from a successful Austrian offensive. After a relatively short bombardment along the whole of the South-Western Front the Russian 8th Army attacked on 4 June 1916, meeting with unparalleled success. The Austrian defences caved in, troops being captured in their dugouts by the hundred, and Brusilov's

The 5th Battery of an unknown artillery unit during the winter of 1916–17. The two gunners nearest to the camera are wearing the pre-war dark green uniform, which was issued at times of shortage. The others are dressed for cold weather in greatcoats and fleece caps. The piece is the M1902 7.62cm Putilov, which was the standard Russian field gun during the First World War. By this date the shell shortages of 1914–15 had been largely overcome, and the artillery was able to enhance its excellent reputation.

The caption to this postcard, dated 1916, is a greeting to the 1st Royal Scots from the 1st Caucasian Rifle Regiment. The crewman of this 90mm mortar, solidly emplaced in a timber-revetted pit, is checking the projectiles.

OPPOSITE **Taken early in 1917, this picture shows a 122mm M1909 Schneider-Putilov howitzer of the 32nd Mortar Half-Regiment** *(Divizion)*. **The gun appears to be brand new. The bleak landscape, and the winter dress of the crew, give a good impression of winter conditions on the barren plains of the Eastern Front.**

men rolled forward. By 12 June nearly 200,000 officers and men, 216 guns and 645 machine guns had been captured and the line had been pushed forward by several miles.

The main Russian summer offensive had been planned for the northern part of the front, opposite the German line. However, after a colossal bombardment and some initial success that offensive came to a halt in the middle of July after just two weeks. In July Brusilov's men again advanced, but by this time they were running into supply problems as the region's transport network was poorly developed. Russian casualties mounted and, in a series of hideously miscalculated attacks employing the methods which Brusilov had specifically abandoned as worthless, the Guards Army was decimated near Kovel. August 1916 witnessed further Russian success on the South-Western Front, but by now the Central Powers were bringing in reserves from all fronts, including two Turkish corps.

Brusilov's system of methodical preparation had achieved advances of up to 50 miles by the time it was called off in the early autumn – an effort unsurpassed by the Western Powers until the summer of 1918.

Training facilities

Russian wartime training facilities were poor. Large numbers of men were kept in the depots in large cities until they were sent to the front as replacements. The training they received was little more than drill and parade ground manoeuvres, due to the lack of instructors with combat experience. With time on their hands, these men were a prey to the political agitation and rumour-mongering which were rampant from the early months of the war; and it was they who supported the revolutionaries of March 1917.

More worthwhile training was carried out at the front, but this depended very much on the resourcefulness of local commanders. Schools were established to train, in a shortened version of the peacetime programme, potential officers who were called up during the war; casualties among the trained, professional officer corps were huge and suitable replacements were difficult to find.

WEAPONS

Small arms

The standard infantry rifle in use was the Mosin-Nagant M1891 7.62mm, a bolt-action weapon with a fixed five-round magazine. Variations on the basic infantry model were the *dragoonskaya* ('dragoon'), which was shorter and lighter, and the *kazachya* ('Cossack'), which was the dragoon type without a bayonet. Artillerymen were issued with the Mosin M1908 carbine.

It rapidly became clear that domestic production would be inadequate, resulting in the importation of American rifles from Westinghouse, Springfield and Winchester; however, the majority of these orders were not filled until 1916–17. Much use was made of captured Austro-Hungarian Mannlicher M1895 8mm rifles, and a factory was turned over to the production of suitable ammunition. Japanese, Italian and French rifles were also issued, along with Berdans from the wars of the late 19th century. The period of acute shortage was 1915 following the 'Great Retreat' from Poland and Galicia, when many weapons were lost. The cruciform-section M1891 bayonet, 17ins. long, was always carried fixed when in the line, the scabbard being left at the depot.

Pistols were issued to artillerymen as well as to officers, the standard model being the seven-shot Nagant M1895 of 7.62mm calibre; but private weapons such as the Colt M1911, Smith & Wesson or Borchardt-Luger were allowed. The Nagant was produced in two versions, double-action for officers and single-action for NCOs and other ranks, as it was believed that the enlisted men would waste ammunition if issued with the double-action model… Captured pieces such as the German 'broomhandle' M1896 7.63mm Mauser, with its wooden holster-stock, were very popular.

A long, curved knife, the *bebout*, was carried by scouts, machine gunners and artillerymen. Caucasian Cossacks had their own long, straight, two-edged knife, the *kinjal*. Naval and Air Service officers wore the *kortik*, a dirk, in place of the sword. The standard sword for mounted troops was the M1881 Dragoon pattern *shashka* sabre; the Cossack variant had no hand-guard and a distinctively shaped grip. Dismounted officers were issued with the 1909 pattern officers' sabre. Other than the Caucasian Cossacks, half of every cavalry regiment was armed with the lance as well as the sword and rifle. The M1910 lance was a hollow steel tube some three metres in length and painted khaki; no pennons were flown in the field.

The standard Russian machine gun was the classic belt-fed M1910 Maxim 7.62mm, originally mounted on a tripod but soon transferred to the more familiar two-wheeled Sokolov carriage, with or without its curved-top armour shield.

Artillery

The mainstay of the Tsar's field and horse artillery was the M1902 7.62cm field gun manufactured by Putilov. The Schneider 7.62cm mountain gun equipped the mountain and horse-mountain units and also replaced the field gun in some army corps.

Medium to heavy guns came in a variety of sizes from a catalogue of manufacturers, and included the

As the fighting settled into the routine of trench warfare the need for close-support weapons became apparent. The Russians developed two types of trench mortar, both of which are seen here with the 'Special Bombing Group' of the 295th Rifle Regt near Stanislau on the South-Western Front. The two mortars in the centre are Likhonin 47mm M1915 on their trench bases – broad chunks of wood to spread recoil and prevent them from sinking into the mud. Both types were man-portable, with a range of some 500 metres. The others are Russian copies of Austrian models.

British M1904 12.7cm (5in., 60pdr), Krupp's 15.2cm M1910 howitzer, and the Schneider M1910 15.2cm howitzer and gun. The Schneider and Krupp weapons were produced under licence in Russia.

Wartime innovations

Trench warfare created the need for small calibre, man-portable guns that could be used in the front line. In 1915 a 37mm trench gun was developed and by 1917 it was planned to equip every infantry regiment with a four-gun battery of these weapons.

The first trench mortars were copied from captured Austrian and German types. However, the most widely used was the robustly simple 58mm M1915 Likhonin. In time for the Kerenski Offensive of 1917 the British provided '60 tons of 2in. mortars, Stokes guns (3-pdr trench pieces) and ammunition'. These weapons were operated by men of Locker-Lampson's RNAS armoured car force, who also acted as instructors in their use.

Machine guns were imported in large numbers, including the 8mm French Hotchkiss M1914, the .303in. Lewis M1915, and particularly the .30cal Colt M1895 'potato-digger'; indeed, so widespread was the use of the latter that the units issued with it were simply listed as *Kolta* detachments.

Armoured fighting vehicles

Russia had experimented successfully with **armoured cars** before the outbreak of war. In 1914 suitable touring cars and lorries were armed and armoured by the firm of Putilov, and the 1st Automobile Machine Gun Battery was created. The success of these units led to the expansion of armoured car companies, until in 1916 the entire force became known as the Armoured Division, with companies allocated to all Fronts. Lacking a domestic vehicle industry Russia imported chassis, notably from Austin, and armed and armoured them herself.

A typical armoured car platoon consisted of two or three twin-turreted machine gun cars such as Austins or Fiats, and one or two armoured lorries – usually American Garfords – mounting short-barrelled 75mm guns. American-manufactured Indian motorcycles provided communications, and some had Maxim machine guns mounted to give anti-aircraft cover. By November 1917 over 200 armoured cars were in service. (Interestingly, the Red Army made the first use of half-tracks as fighting vehicles in 1920, using converted Austins.)

Estimates vary as to the number of **armoured trains** Russia possessed in 1914, between two and ten. At first the guns

Russia maintained the largest force of armoured cars of any of the Allies; the diversity of types included this Fiat. Strangely, the tyres are of the studded KT type, which were intended for driving on snow. Armament and armour were usually added to the various basic foreign chassis in Russia, and this twin-turret superstructure was popular. The two members of the crew are both wearing the issue leather suits with the armoured troop's cap – see Plate F1.

The use of armoured trains was widespread on the Eastern Front. The Russian officer at left, wearing the 'duster' coat and goggles often used for travelling by car, is inspecting the shell-battered remains of a typical Austrian train. By March 1916 armoured trains were organised by weight of artillery into 'shock' (7.62cm), 'fire support' (107mm and 122mm naval), and 'heavy fire support' (152mm and 203mm naval) groups. All were armed with plentiful machine guns to provide close defence against infantry or cavalry. Men of the railway battalions crewed the trains, sometimes with naval gunners attached.

lacked traverse, but turrets were introduced to compensate for this, and machine guns to ward off attackers. However, there was no standard design, and armament was a case of 'as much as possible, of whatever is available'. Armoured trains came into their own with the advent of positional warfare, through their ability to provide mobile artillery support despite the lack of metalled roads in much of the area of operations.

The vast majority of Russia's **aircraft** in 1914 were of French manufacture – Nieuports, Morane Type Gs and Deperdussins, plus some Farmans. Aircraft imported during the war included the Spad VII and Nieuport 17. The navy operated Curtiss Type D and E floatplanes, Type F flying boats and Sikorski S-10 floatplanes. Captured aircraft pressed into service included Aviatiks, Albatrosses and LVGs. Many were re-armed with the Colt machine gun, which was very popular with Russian pilots.

SELECT BIBLIOGRAPHY

Relevant English language titles are as follows:
Allan, W.E.D. & Paul Muratoff, *Caucasian Battlefields,* (r/p 1999)
Bruce-Lincoln, W., *Passage through Armageddon,* (Oxford 1986)
Figes, Orlando, *A People's Tragedy,* (London 1996)
Golovine, Gen.Nicholas N., *The Russian Army in the World War,* (New Haven 1931)
Knox, Maj.Gen.Sir Alfred, *With the Russian Army 1914–17,* (London 1921)
Mollo, Andrew, *Army Uniforms of World War 1,* (Poole 1977)
Stone, Norman, *The Eastern Front 1914–1917,* (London 1975)
(War Office), *Handbook of the Russian Army 1914,* (r/p 1996)
Wildman, Allan K., *The End of the Russian Imperial Army,* 2 vols (New Jersey 1987)

Aircraft such as this Nieuport 17c scout from France reached the Russian armed forces via Vladivostock and Archangel, albeit slowly. This machine appears to be factory fresh, with only the national roundels – red, blue and white, reading inwards – added on the underside of both wings. Britain and France provided instructors to familiarise Russian pilots and ground crew with imported machines.

THE PLATES

A: STAFF

A1: Captain of the General Staff A.Afanasyev, 1917

The typical officer's uniform of the later war. The tunic was known as the 'French'; his breeches and boots are standard patterns. Note that the shoulder boards are now of unstiffened material, and are sewn into the tunic seams to make them less prominent. Staff aiguillettes were either white or khaki. He is decorated with the Order of St Anne 2nd Class with Swords, the Order of St Stanislaus 2nd Class with Swords, and the Order of St Vladimir 4th Class with Swords, and wears on his right breast pocket the badge of the Nikolaevski Military Academy; he also carries a gilded St George's Sword 'for Bravery'. All these figures wear the more elaborate officers' version of the cap cockade, of domed construction with the rayed edge making a crenellated effect.

A2: General of Artillery V.Irmanov, 1914

This illustrates the officers' greatcoat, in this case with the scarlet lapels and cuff piping which distinguished general officers; his rank is also shown by his shoulder boards, in the zig-zag patterned lace of general officers. The black and red details on the service cap and collar patches indicate the arm of service. He wears the Order of St George 3rd Class and the ribbon of the 4th Class. Hidden here by the coat, his breeches would have scarlet stripes – *lampasi*. General Irmanov commanded III Caucasian Corps in 1914. Following the March 1917 Revolution he resigned his commission and re-enlisted as a private soldier.

A3: Adjudant General of Cavalry A.A.Brusilov, 1916

Russia's most effective general of the war wears a plain, regulation officer's service uniform with staff aiguillettes, the general officer's trouser stripes being the only concession to colour. He displays the Order of St George 3rd and 4th Class, and the badge of the Corps of Pages, and carries a gilded St George's Sword for Bravery.

A4: Colonel, medical service, 1915–16

The doctor is wearing the *Bekesha* which was a popular alternative to the usual greatcoat; this was a long version of the fleece-lined *polushubok* worn by Plate B4. His earmuffs are privately purchased. All members of the medical services wore the red cross brassard on the left arm.

B: GUARDS

B1: Staff Captain, Grodno Life Guard Hussar Regiment, 1915

The crimson hussar breeches were commonly seen at this early stage of the war. The boots are of the style worn by hussars throughout Europe, high-fronted and with a white metal rosette at the front. Note the whistle on his left shoulder brace. On the breast pockets he wears the badges of his regiment and the Officer's Cavalry School.

B2: Private, Semenovski Life Guard Infantry Regiment, 1914

In full summer service dress, this Guardsman typifies what the British military attaché described as 'the finest human animals in Europe'. On the cuff of his *gymnastiorka* can be seen the white lace identifying the 1st Guards Infantry Division, and on the chest and shoulder straps the blue lace of the division's second regiment. The Guards were the only infantry to be issued knapsacks for wear on the back. The ends of his rolled greatcoat are tucked into his mess-tin, a common practice.

B3: Cossack, Ataman's (Tsarevitch's) Life Guard Cossack Regiment, 1917

Cossacks always wore their headgear in defiance of gravity,

This typically posed studio shot of *Pod-porutchik* (2nd Lt) Mikhail Zheleznyak shows the officer's field equipment for the campaigns of 1914–15; he was killed in October 1914. Oddly enough, the wearing of the shoulder braces vertical at the front and crossed at the back was banned, along with officers' shoulder boards, after the November Revolution of 1917. The M1881 sword is slung in the 'Oriental' style, edge upwards; the pistol on the right hip is probably an M1895 Nagant. During the early part of the war shoulder straps were often worn with the decorated side uppermost; for officers the lace colour corresponded with that of the regimental buttons, either gold or silver. Cf Plates B & C.

This soldier wearing his greatcoat – fastened by hooks-and-eyes and with a row of brass buttons for decoration only – shows his volunteer status by the twist piping on his shoulder straps in white, orange and black. The collar patches on the greatcoat were in the arm-of-service colour, sometimes with appropriate piping. The service cap is worn in the Russian style, pushed back on the head. The peak is black but was usually painted green in the field; dismounted troops did not have chinstraps; and the metal cockade was in Romanov colours for all enlisted men.

hence the rakish angle of the fleece cap. Longer hair, with a lovelock just visible, was also a token of the Cossack's freebooting past. The final Cossack touch is the slinging of the rifle over the right shoulder, rather than the usual left. Note the great length of the cavalry coat, and the deep pointed cut of the cuff, front and back. White leather waist and sword belts show that this is a Guardsman. The M1910 lance has no pennon; these were not used in the field.

B4: Volunteer, Her Majesty's Life Guard Lancer Regiment, 1916

This class of short term volunteers provided many of the reserve officers called up at mobilisation. The *Romanovski polushubok* was a popular winter jacket of fleece-lined cloth; note the fancy 'lancer plastron' effect on the chest of this example. 'Volunteer braid', of twisted orange, white and black, edges the shoulder straps. These bear a typical regimental monogram – here that of the Dowager Empress Maria Fedorovna, the regiment's Colonel-in-Chief, which was only worn by the Elite Squadron. By this stage of the war much of the Russian cavalry was out of the line waiting for a breakthrough, hence the blue breeches and lack of weapons.

C: INFANTRY
C1: Lieutenant-Colonel, 94th Yeniseiski Regiment, 1914

A rather crumpled 'duster' coat is worn to protect the uniform; this *pod-polkovnik* has attached his shoulder boards of rank to this, identified by two stripes and three stars. Wearing medals was obligatory, even in combat; this officer has the Order of St Anne 3rd Class with Swords. As well as the pistol holster and binocular case he has a torch at his belt.

C2: Private, 404th Kamyshinskiy Regiment (Opolchenie), 1915

From his service cap to his boots this man typifies a *frontovik*. Above the Romanov cockade on his cap is the 'militia cross' of the Opolchenie bearing the Tsar's monogram (non-Christian militia wore an octagonal badge instead). The coloured side of the shoulder straps was red with yellow paint stencilling in both the regular and militia infantry; the 404th Regt was made up of the 214th, 215th and 216th Bns from Saratov. His medal is for the Russo-Japanese campaign in which he fought as a young regular. To supplement the pouches on his waistbelt he has slung two canvas bandoliers, each holding 60 rounds, over his shoulders. Hanging at his left hip is the 'bread bag', the large haversack that did duty as a knapsack for non-Guard infantry; officially it was supposed to accommodate two shirts, a pair of drawers, two pairs of foot clothes (worn by Russians in place of socks), towel, mittens, sewing kit, 4½lbs of biscuit, salt, rifle cleaning kit, and cup.

C3: Senior NCO, regimental machine gun kommando, 8th Moscow Grenadier Regt, 1917

The three lines of red lace on his shoulder straps identify a *starshi-unteroficer* (roughly, sergeant); the button on his collar patches, an NCO; and the crimson cuff lace indicates that he is a member of the machine gun team. From August 1914 'M' for Moscow on the shoulder straps replaced the monogram of the regiment's Colonel-in-Chief, Duke Frederick of Mecklenburg. This NCO is turned out in regulation winter dress, and armed with a pistol and *bebout* knife. The ends of the *bashlyk* slip under the shoulder straps and tuck into the belt – note the brass Grenadier plate.

C4: Senior private, grenadier platoon, 4th Rifle Brigade ('The Iron Brigade'), 1916

Wearing the *pilotka* forage cap, this *yefreitor* displays his rank by the red transverse lace stripe on his shoulder straps. On the upper left sleeve is a flaming grenade badge, the mark of these picked assault platoons. In 1916 wound stripes were introduced on the left cuff; for other ranks and NCOs these were in red, for officers in gold or silver depending on the regimental button colour. The armament reflects his training for close combat and infiltration: a hatchet, a carbine and a bag of grenades. He also has a gas respirator slung on his left hip. Knee boots were frequently replaced with ankle boots and puttees by this stage of the war.

D: LINE CAVALRY
D1: Bombardier Layer, 20th Horse Artillery, 1915

Very little, other than his shoulder boards, distinguishes this bombardier from any other mounted soldier. The sword and pistol are standard issue. To his left is a Danish manufactured M1904 Madsen light machine gun of the type issued to some cavalry regiments before 1914.

D2: Captain, 5th Alexandriyski Hussar Regiment ('Immortal Hussars'), 1916–17

The unit was known as the 'Immortal Hussars' after a line in its regimental song. This officer is very much in the wartime fashion with his black leather jacket cut like the uniform tunic, gloves, and regimental breeches striped with silver lace. The brown leather equipment is regulation issue of the 1912 pattern. Again, note the hussar boots with rosettes.

D3: Trooper, 16th Tverskoi Dragoon Regiment, 1915–17

The 16th Dragoons served with the Army of the Caucasus. This trooper wears a fleece cap, cowl and gloves in cool weather; his greatcoat is strapped to the front of the saddle. The horse furniture is the standard issue brown leather with saddlebags at the rear. Regular cavalry carried the bayonet, which can be seen attached to the sword scabbard in the '1881 method'.

E: COSSACKS

E1: Cossack, 1st Argun Regiment, Trans-Baikal Host

Cossacks were expected to provide their own uniforms; and this man shows the regulation dress for the Steppe Hosts, distinguished by the yellow trouser stripe of the Trans-Baikal troops. His weapon is the 'Cossack' model of the standard M1891 Mosin-Nagant '3-line' rifle.

E2: Cossack, Kuban Cossack infantry

Kuban infantry units wore the traditional Caucasian Cossack dress including the heavy black felted hair *burka* cape for foul weather. His weapons include rifle, pistol, and a Caucasian dagger which has its highly decorated scabbard protected by cloth bindings. The standard shirt-tunic is worn under a *kaftan* coat. Kuban and Terek Cossacks shaved their heads.

E3: Lieutenant, 2nd Volgski Regiment, Terek Host

The epitome of the Caucasian Cossack officer; the highly decorated weapons and *kaftan* are typical of these units throughout the war. The cartridge pockets on each breast, *gaziri*, were functional as well as decorative. The undershirt, *beshmet*, was often privately made and did not always conform to regulations. During the war supply problems led to khaki replacing the grey *kaftans*. The rank of this *sotnik* or first lieutenant is identified by the three stars and single stripe on his shoulder boards, which also bear the regimental number '2' and the Cyrillic initial of the Terek Cossacks, which resembles 'Br'. Light blue was the traditional distinguishing colour of the Terek Host. He wears the Order of Vladimir 4th Class with Swords, the Order of St Anne 4th Class with Swords, a Terek Cossack badge and that of the Novocherkask Cossack School. His handsome weapon is a St Anne's Sword 'for Bravery' – note the rosette in the pommel. He carries the Cossack *nagaika* whip.

E4: Warrant Officer, 17th Don Cossack Regiment General Baklanov, 1914

The 17th Don Regt wore on their caps the scroll and death's-head as a reminder of past glories; they were named for Gen. Yakov Baklanov, a hero of the Crimean War. This veteran *pod-khorunji*'s extraordinary collection of decorations testifies to his personal bravery and skill at arms: 1st, 2nd, 3rd and 4th Classes of the St George's Cross, St George's Medal 4th Class, Russo-Japanese War medal, 300th Anniversary of the Romanov Dynasty, 1st, 2nd and 3rd Classes of the Sharpshooter's Badge, and the crossed swords marking proficiency at swordsmanship. His jacket is a non-regulation *nagolny polushbok* cut to a practical length for riding; it is left in natural sheepskin rather than being covered with fabric. Note the *bashlyk* bundled around his neck.

F: SPECIAL TROOPS

F1: Armoured car driver, 7th Automobile Machine Gun Platoon, 1915

The 'Swedish *kurtka*' leather jacket, gauntlets and breeches were issued from the beginning of the war. This style of

An evocative shot of an infantry officer amid the debris of war. The khaki 1917 pattern shirt-tunic has the soft shoulder straps stitched into the shoulders. The breeches are dark green; and note that the boots come above the knee. He wears the officer's Order of St George in white enamel, and either his regimental or a military school badge on his left breast. The coloured parade cap is set well back; and his sword has been replaced with a walking stick.

Three Don Cossack cavalrymen photographed out of the line, wearing the *gymnastiorka* and (centre) the 1912 *kittel* tunic. The Cossack sword held by each man is slung by the belt over the right shoulder. The Cossack on the right is a lance-corporal – note the transverse lace on his shoulder strap; the broad breeches stripe, just visible, is in the red of the Don Host. Cf Plate E1.

M1911 cap was unique to the motorised branch of the Engineers from whom they had developed; a flatter, square-peaked cap (see Plate F5) was also worn by enlisted men. Fold-down padded earflaps cut down the noise inside the vehicle. Officers wore dark green breeches striped with red. Shoulder straps and other insignia were worn as usual on the leather jacket; the number '7' and the branch-of-service badge on the shoulder straps identifies his unit. Armament was limited to a revolver for practical reasons.

F2: Stretcher bearer, 1915–17
The prominent red crosses on his brassard, service cap, shoulder straps and medical bag show this man's branch of service clearly. Note the method of carrying the stretcher by the handles, not the poles. The breeches are grey-blue with light blue piping. The tunic is the M1910 pattern.

F3: Regimental Orthodox priest
In Russia religion played a significant part in people's lives and this carried over into the armed forces. The regimental priests, known as *Batyushka*, were a common sight at the front, and ceremonies were observed punctiliously. The blessing of the troops before battle was often carried out by priests in their full regalia. The priest here is wearing the traditional overcoat and hat of a cleric over his everyday cassock; the *Naperstny* cross is his badge of office, hanging from a St George's ribbon to mark his bravery under fire.

F4: Pilot officer, Aviation Service, 1914
Clad in leather flying clothing, this pilot is armed with the Mauser M1896 pistol favoured by many Russian officers. On his chest is the speaking tube with which he communicates with the observer. The flying helmet – note its cockade – was initially imported from France. The breeches and jacket collar are piped in red.

F5: Cyclist, 3rd Bicycle Company, 1915–17
Cyclists seem out of place in Russia, which boasted few metalled roads. Originally armed only with pistols, they were issued during the war with Japanese Arisaka rifles. During the last year of the war cyclist battalions were formed, and were noted for their bravery during the retreat into Russia following the collapse of the 1917 summer offensive. Their uniform and equipment was designed for practicality, hence the puttees, canvas ammunition belt, goggles, and flatter service cap with a larger peak. On his left sleeve is a wound stripe; he wears the St George's Medal, and on his right brace a 3rd Class Sharpshooter's Badge. The cycle could be folded and slung on the back.

G: NATIONAL TROOPS

G1: Private, 5th Latvian Rifle Regiment, 1916–17
Armed with the .30cal Winchester M1895 rifle, this private is wearing the canvas equipment issued during the war when leather became scarce. The copy of the French Adrian steel helmet was manufactured in Finnish factories mainly in this version for Latvian and Czech troops, without holes at the front for the attachment of a badge.

G2: Trooper, Turkmen Horse Half-Regiment, 1914–15
This was a two-squadron half-regiment voluntarily recruited among the Moslem Tekin tribe of Turkestan. Over the regulation shirt-tunic the men wore a yellow and orange striped *kaftan* based on their native dress, with yellow shoulder straps. They also wore a white *bashlyk* cowl piped in pale blue, and black breeches. Officers' service caps were black with a yellow band and crown piping; their breeches were blue, striped yellow. The fleece cap worn by all non-commissioned ranks throughout the year was this more shaggy, Asiatic type. As Gen.Kornilov's bodyguard during 1917, they accompanied him everywhere.

G3: Trooper, 'Savage Division', 1914–17
The Caucasian Native Cavalry Division served on the Southern, South-Western and Romanian fronts. All regiments are supposed to have worn the *cherkesska* coat, the individual units being distinguished by the colour of the shoulder straps and cowls, the latter being piped with white: Daghestan Regt – light blue straps, red cowls; Kabardian Regt – light blue straps, white cowls; Tartar Regt – red straps, burgundy red cowls; Chechen Regt – light blue straps, yellow cowls; Circassian Regt – red straps, white cowls; Ingush Regt – red straps, light blue cowls.

One source states that all units wore black *kaftans*; others, that the Ingush Regt had shoulder straps piped in white, red and blue twist. As volunteers, all ranks probably enjoyed considerable latitude (e.g. this rider wears a privately made *kaftan* with fleece trim), and all descriptions may well be correct. The majority of the officers were Russians, as were the technical troops and gunners. The machine gun sections were sailors from the fleet, who wore Circassian dress with naval shoulder straps.

G4: Lieutenant, Polish Lancers, 1917
The details on this officer's uniform mark him as Polish: the white Polish eagle on the cap, the crimson breeches stripes and cuff lace, and the eagle badge on the breast pocket.

H: ELITE UNITS 1917–18

H1: Lieutenant, Shock Battalion of Rear Echelon Volunteers, 1917
The movement to continue the war after the March Revolution produced various schemes, including the idea of forming 'Shock Revolutionary battalions from rear volunteers'. A good number came forward, mainly from cadet schools and reserve units. Cuff ring rank insignia were introduced during the brief period of the Provisional Government as an alternative to the shoulder straps, which had Tsarist overtones. The badge on his right arm was issued to all ranks of these units and was worn on both tunic and greatcoat; and note the death's-head on the shoulder straps. A piece of black and red cloth was often substituted for the Imperial cockade. This officer wears a soldier's grade St George's Cross awarded by his men in recognition of his bravery; he also displays a university badge.

H2: Senior NCO, 1st (Kornilov's) Shock Regiment, 1917
This sergeant wears the shoulder straps and left sleeve badge of the famous Kornilov regiment. On his right arm is the chevron of the Shock units, red and black symbolising revolution and death. Note the death's-head helmet emblem in cast metal, although these were often painted on in white. The white belt and cuff piping show that this man was formerly in the 1st Guards Division.

H3: Private, Russian 'Legion of Honour'; France, 1917–18
The Russians sent infantry units to serve on both the Western and Salonika Fronts in token of their solidarity with the Western Allies. In both cases the French provided clothing (here, in Colonial khaki) and equipment; these were worn with Russian shoulder straps and, in the case of the troops in France, a cloth Russian tricolour on the left sleeve of the tunic or greatcoat. The 'LR' on the helmet and collar patches stands for *Légion Russe*.

H4: Private, 1st Women's Death Battalion, 1917
The 1st Women's Battalion of Death was raised in Petrograd during May 1917 at the instigation of Sgt. (later Lt.) Maria Botchkareva who, after serving at the front since 1915 and being disgusted by the condition of the army, was granted permission to form this unit. Accorded the unique honour of carrying a standard featuring Botchkareva's name, the battalion, numbering 1,000 women, took part in the Kerenski Offensive of summer 1917, sustaining heavy casualties. Other women's battalions were formed, at least one in Moscow and another glorying in the name of 'The Black Hussars of Death'. The uniform differed little from that of the men other than the shoulder straps and the shock unit chevron (note black-and-red stripe on the former).

H5: Private, 2nd Voluntary Detachment of Crippled Warriors, 1917
This strangely titled unit was raised in early summer 1917 from volunteers who had been severely wounded and were now based in Petrograd, organised into two battalions of some 1,000 officers and men. The Cyrillic letters in the sleeve chevron are the initials of the words 'shock' or 'storm detachment'. He wears ribbons of the Order of St Vladimir, and the regimental badge of the 69th Ryazanski Infantry Regiment. His sidearm is a regulation issue sword.

(Details:)
1A: Private's shoulder strap, Shock Battalion of Rear Volunteers, 1917
5A: Cuff title, Committee for Recruiting Disabled Soldiers, 1917
6: Sailor's shoulder strap, Reval Naval Shock Battalion, 1917
7: Private's shoulder strap, St George's Battalion, STAVKA 1916–17
8: Private's shoulder strap, 1st St George's Battalion, Kiev, 1917
9: Cap badge, Death Battalions, 1917
10: Cap badge, Shock Battalions, 1917

Taken in Romania during May 1917, this photo shows two Aviation Service observers dressed in leather jackets and trousers, with ankle boots and gaiters. One wears the coloured service cap, the other the khaki version; such combinations were not uncommon. Black leather breeches and jackets were very popular with young officers during the last year of the war. Cf Plate F4.

INDEX

Figures in **bold** refer to illustrations

aerial reconnaissance 38
Air Service. *see* Aviation Service
aircraft **15**, **42**, 42
Alexeyev, General M.V. 8
'Alien' troops (*Inorodtsi*) **G**, 11, 12, 14, 20, 46-47
armoured cars 19, 41, **41**
 drivers **F1**, 45-46
Army of the Caucasus 4, 13
artillery 12, 13, **13**, 14-15, 24, 36, 37, 38, **38**, 40-41
 Cossacks 14
 fortress 4, 15
 Guard 16
 observation post **6**
 Savage Division 18-19
 small arms 39, 40
 uniforms **D1**, 23, 44-45
Austria-Hungary 4, 5, 38-39
Aviation Service **F4**, **15**, 19, **20**, 40, 46, **47**

Belgian troops 19
bibliography 42
Bolsheviks, the 10, 11, 18, 20
Brusilov, General A.A. **A3**, 8, 9, 38, 43
Brusilov Offensive, the 9, 16, 38-39

caps **11**, 21, 24, 33, **44**
 fleece **10**, **22**, 33, **37**
Caucasian Cavalry Division 13
Caucasian Grenadier Division 12
Caucasian Native Cavalry Division **G3**, 14, 18-19, 46-47
Caucasian Rifle Brigades 13, 14
cavalry 13, 15, 23, 33, 36-37
 distinctions 22(table)
 Guard **B1**, **B4**, 16, 43, 44
 irregular **G2-3**, 12, 18, 23, 46-47
 uniforms **D**, 44-45
 weapons 39, 40
chronology 5-11
cockades 33, **44**
collar patches 24, 24(table)
conscription 3, 11
Cossacks 11, 12, **12**, 13, 13-14, 15, 19, 33, 37, 40
 distinctions 23(table)
 flags 35-36
 Guard **B3**, 16, 43-44
 infantry **E2**, 13, 45
 tactics 36-37
 uniforms **E**, 21, 23, 45, **46**
cyclists **F5**, 46
Czech Legion, the 21

Daghestan Native Cavalry Regiment 14
Death Battalions **H9**, 17-18, **18**
distinctions 22(table), 23, 23(table), 24(table), 33(table)
Don Cossacks **E4**, 13, 14, 16, 19, 45, **46**

East Prussia 4, 5
Eastern Front 5
Eastern Front 1914-17, The (Stone) 3
engineers 15, **23**
equipment 33-34, **37**, **43**
exemptions 11

Finland 11
Finland Rifle Brigades 13, 14
'Finnish' units 11
flags 35-36
France and French troops 4, 15, 19

gas masks 36
Germany 4
Great Britain and British troops 4, 15, 19-20
greatcoats **A2**, **10**, **11**, 23-24, **34**, 43, **44**
Grenadier Corps 12, 14, 17, 24
grenadier platoons **C4**, 17, 34, 37, 44
Guard, the 12, 13, 14, 15-17, 24, 33, 34
 cavalry **B1**, **B4**, 5, 13, 16, 43, 44
 distinctions 22(table)
 engineer battalion 15
 uniforms **B**, 22, 23, 43-44

helmets 34

image 3
Imperial Militia (*Opolchenie*) 11, 12, 21, 33
infantry 12-13, 14, 15, 35, 36, **37**, 38
 Guard **B2**, 15-16, 43
 uniforms **C**, **E2**, 3, **21**, 44, 45
Irmanov, General V. **A2**, 7, 43

Jaroslavice, battle of, 20 August 1914 5, 37

Kerenski Offensive, the 10, 19, 47
Kornilov, General L.G. 10, 18
Kuban Cossacks **E2**, 13, 14, 16, 33, 45

Latvia and Latvian troops **G1**, 20, 46
Lenin, Vladimir Ilyich (1870-1924) 10

medals and decorations 17, **20**
medical services **A4**, **F2**, 7, 43, 46
military districts 11
mobilisation 4, 4-5, 5, 13
Moslems 11, 12, 14, 18
munitions production 3

naval personnel 16, 18, 19, **19**, 40
Nicholas, Grand Duke 3, **4**
Nicholas II, Tsar (1868-1918) 3
Non-commissioned officers **C3**, 11, 24, 40, 44

officers **H1**, 3, 9, 19, 24, 33, 39, 40, **42**, **43**, 47
 Cossacks **E3**, 45
 shoulder straps 35(table)
 uniforms **A1**, 17, 21-22, 23, 24, 34, 43, **45**
Orenburg Cossacks 14, 35-36
organisation
 artillery 14-15
 cavalry 13, 15
 Cossacks 13-14, 15
 infantry 12-13, 15
 Storm Battalions 18
 technical branches 15
 wartime innovations 15
Ossetian Cavalry *Division* 14
Ottoman Turks 4

partisans 37
Plan 19 4-5
Poland and Polish troops **G4**, 20, 47
priests **F3**, 46

railways 4
ranks, insignia of 23, 34, 35(table)
recruits and recruitment 3, 11, 17, 20
reforms 3
Revolution, the (1917) 10-11, 16-17, 34, 39, 47
Rifles, the 12-13, 14, 15, 16
Russia and the Russian empire 3, 3-4
Russian Armoured Car Division, Royal Naval Air Service 19
Russo-Japanese War (1904-5) 3, 18, 21, 36

sappers 12
Savage Division **G3**, 14, 18-19, 46-47
Serbian troops 20
Shock Battalions **H**, 47
shoulder boards 22-23, 34
shoulder straps **H**, 21, 23, 24(table), 33(table), 34, 35(table), 47
Siberian rifle regiments 12, 13, 14, **14**, **22**, 33
Special Army, the 16, 17
St George's Battalions 17
Staff officers **A**, 34, 43
Standing Army 11
Stone, Norman 3
Storm Battalions 17-18
strength 11
subject races **G**, 11, 12, 14, 20, 46-47
Sukhomlinov, General V.A. 3

tactics 36-37, 38-39
technical branches 15, 24, 33
Terek Cossacks **E3**, 13, 16, 45
terms of service 11-12
training 39
trains, armoured 41-42, **42**
Trotsky, Leon (1879-1940) 11
Turcoman Cavalry *Division* **G2**, 14, 46
Turkestan Rifle Brigades 13, 14

uniforms **10**, 21, 21-22, 23, 34, **34**
 artillery **D1**, 44-45
 cavalry **D**, 44-45
 Cossacks **E**, 23, 45, **46**
 Guard **B**, 22, 23, 43-44
 infantry **C**, 3, **21**, 44
 officers **A1**, 17, 21-22, 23, 24, 34, 43, **45**

Voluntary Detachment of Crippled Warriors **H5**, 47

weapons 15, 40
 machine guns **D1**, 12, 13, 15, 36, 40, 41, 45
 mortars **39**, **40**, 41
 rifles **37**, 39-40
 wartime innovations 41
weather 5
western defences 4
Western Front **H3**, 47
women **H4**, 14, 18, 18, 47
Women's Battalion of Death **H4**, 18, 47

COMPANION SERIES FROM OSPREY

ESSENTIAL HISTORIES
Concise studies of the motives, methods and repercussions of human conflict, spanning history from ancient times to the present day. Each volume studies one major war or arena of war, providing an indispensable guide to the fighting itself, the people involved, and its lasting impact on the world around it.

CAMPAIGN
Accounts of history's greatest conflicts, detailing the command strategies, tactics, movements and actions of the opposing forces throughout the crucial stages of each campaign. Full-colour battle scenes, 3-dimensional 'bird's-eye views', photographs and battle maps guide the reader through each engagement from its origins to its conclusion.

ORDER OF BATTLE
The greatest battles in history, featuring unit-by-unit examinations of the troops and their movements as well as analysis of the commanders' original objectives and actual achievements. Colour maps including a large fold-out base map, organisational diagrams and photographs help the reader to trace the course of the fighting in unprecedented detail.

ELITE
This series focuses on uniforms, equipment, insignia and unit histories in the same way as Men-at-Arms but in more extended treatments of larger subjects, also including personalities and techniques of warfare.

NEW VANGUARD
The design, development, operation and history of the machinery of warfare through the ages. Photographs, full-colour artwork and cutaway drawings support detailed examinations of the most significant mechanical innovations in the history of human conflict.

WARRIOR
Insights into the daily lives of history's fighting men and women, past and present, detailing their motivation, training, tactics, weaponry and experiences. Meticulously researched narrative and full-colour artwork, photographs, and scenes of battle and daily life provide detailed accounts of the experiences of combatants through the ages.

AIRCRAFT OF THE ACES
Portraits of the elite pilots of the 20th century's major air campaigns, including unique interviews with surviving aces. Unit listings, scale plans and full-colour artwork combine with the best archival photography available to provide a detailed insight into the experience of war in the air.

COMBAT AIRCRAFT
The world's greatest military aircraft and combat units and their crews, examined in detail. Each exploration of the leading technology, men and machines of aviation history is supported by unit listings and other data, artwork, scale plans, and archival photography.